Project Engineering

Project Engineering

The Essential Toolbox for Young Engineers

Frederick B. Plummer, Jr.

AMSTERDAM • BOSTON • HEIDELBERG • LONDON
NEW YORK • OXFORD • PARIS • SAN DIEGO
SAN FRANCISCO • SINGAPORE • SYDNEY • TOKYO
Butterworth-Heinemann is an imprint of Elsevier

ELSEVIER

Butterworth–Heinemann is an imprint of Elsevier
30 Corporate Drive, Suite 400, Burlington, MA 01803, USA
Linacre House, Jordan Hill, Oxford OX2 8DP, UK

♾ Recognizing the importance of preserving what has been written, Elsevier prints its books on acid-
free paper whenever possible.

Library of Congress Cataloging-in-Publication Data
Application submitted

British Library Cataloguing-in-Publication Data
A catalogue record for this book is available from the British Library.

ISBN: 978-0-7506-8279-4

For information on all Butterworth–Heinemann publications
visit our Web site at www.books.elsevier.com

Transferred to Digital Printing in 2013

Working together to grow
libraries in developing countries

www.elsevier.com | www.bookaid.org | www.sabre.org

ELSEVIER   BOOK AID
            International   Sabre Foundation

Dedication

To my mother and father for teaching me the first lesson of project engineering—integrity;

To Sigrid Melle, my wife and best friend, who got me started on this project, graciously gave me the freedom to pursue it, and has always believed in it;

To my daughter and three sons for their inspiration, and the confidence they have given me in the next generation; and

To the memory of Lesley Summerhayes, editor, colleague, and friend.

Contents

Chapter 3

A Crash Course in Management

Chapter 4

How Projects Work

Chapter 5

Learning Project Engineering on the Job:
A Case Study

Chapter 6

Skills That Can Get You Ahead

Chapter 9

Advice from the Pros

Chapter 10
Approach the Job with Confidence

Preface

When you think of a toolbox, you imagine a collection of handy implements organized well enough that you can find them when you need them. They're durable and collected over time—some handed down from generation to generation. Most of them are essential, for without them you couldn't imagine starting a tough task. If you add to your collection as the need arises, and if you keep them sharpened, rust-free, and maintained, they'll serve you well for a lifetime.

I view this book as a "starter set" of tools for the engineer just arrived at his or her first job, and for the engineer with a few years' experience (I won't draw the line on how many) who wants to make the transition from a technical assignment to a leadership position.

Project engineering is a chance to lead and coordinate the efforts of others on a project or part of a project. Especially in smaller companies, you may be thrown into the middle of a significant task shortly after you've walked in the door and found out where the restrooms are. It's crucial to be prepared and not be caught flatfooted. That's when this book can help you the most.

The book is organized in the order you need the information. The first half presents the fundamentals of project engineering. It begins where you begin when you take on your first job. It then lays out what you have to do as a project engineer in terms of fundamental duties that are common to almost every project engineering situation. A crash course in management gives the essential principles of leadership and the basic skills for getting along with coworkers and management. An overview of how projects work helps you visualize how project engineers fit into the big picture. A case study takes theory into practice for a young project engineer in a realistic situation; it's like a transfusion of practical experience.

Fundamentals are one thing, but success is another. The second half of the book moves to the broader and perhaps more important topics of being successful and managing your career. Being effective in the performance of your job, competing with contemporaries, and dealing with office politics are balanced against the values of ethical business conduct. Tools for working internationally add the cultural skill set needed to compete in our global business world. That's followed by a chapter of advice from project engineers and managers—wise guidance specifically tailored for young project engineers—addressing the points those project pros feel are most important. The book ends with a description of the most powerful asset you can have: confidence.

I hope you will find this unique collection of technical, business, personal effectiveness, interpersonal skill, and leadership tools of value in getting you off to a good start and headed in the right direction on your career journey. The book doesn't cover all aspects of project management; it just explicates what over 30 years of coaching and counseling young engineers have shown me to be essential. These pages are filled with what a friend called high-end common sense.

To keep it short and to the point, I've used the criteria that the material must be *essential*. When you find the book going into more detail on one topic or another, it's because most engineering schools don't teach it, but you need to know it. If you don't need to know it right now, I predict that you will. So mark the passages you find most useful now, and make sure to glance back over the rest of the book a little further along the road.

In addition to making the material essential, I strive to always present *something you can actually do* to solve the challenges confronting you. Use what you need when you need it and seek important training. You won't want to change your entire style overnight, but over time your performance and effectiveness will improve.

Like any good, basic toolbox, I expect you will use the book many times during your early years, and less as time goes by. But even as an accomplished project engineer you may find yourself coming back to put your hands on the basics every so often. And just as good tools get loaned to someone who has a job to do, someday you may lend the book, or advice from it, to other people.

Acknowledgments

A number of people in my writing community have helped give life and energy to this book. I'm indebted to my teachers, Elizabeth Harper Neeld, Alexis Glynn Latner, and Jacqueline Simon, for introducing me to the craft and for their encouragement. Alexis, with her uncanny insights into things technical, went on to be my "ideal reader," editor, mentor, and a continual source of support, advice, and confidence.

My thanks go to Lesley Summerhayes. She put aside her disdain for engineering and pitched in with constructive criticism on the voice, the concept of the case study, and the conception and shaping of those early chapters. Her death was a tragic loss.

So many colleagues and friends have contributed to the content. The list starts with my brother, Ken Plummer, and my sons, Fred Plummer III, Robert Plummer, and Jon Erlend Holand, for their comments, advice, insights, and plain, ordinary help when I needed it. Other key contributors include Lynn Boyd, Mike Brady, Kristin Farry, Jim Flood, Arnt Erik Hansen, Paul Hellen, Rob Howard, Jeff Hulett, Rick Johnson, Jing Kuang, Peter Lacey, Olav Lappegaard, George Lock, Lisa Solberg Mallon, Don Maus, Will Moon, Christina Nordstrom, Yarami Pena, Ruzica Petkovic, Dick Rolstad, Haakon Sannum, Sandy Setliff, Bjorn Solheim, Ray Steinmetz, and Harry Underland. All of these people made their mark on this book in one way or another but three stand out: Jeff Hulett for his priceless collection of project proverbs and a score of other inputs, Yarami Pena for his wise counsel in the formative stages when I was deriving the content and synthesizing the project engineer's duties, and Harry Underland for improving my understanding of manufacturing.

Special mention goes to Eric Namtvedt and his crew at FloaTEC, LLC, for bringing me out of retirement to learn the "contractors' world" and for their permission to publish their artwork. I'm indebted to Malcolm Taylor for his candid insights and Ricky Brown for help with the figures. The many conversations with

Liz Maraist on the subject of business conduct helped me calibrate my understanding of that crucial subject.

Above all, I can't begin to thank those who made possible the transition from book project to the published book: Bud Griffis, Dan Morris, Joel Stein, Jeff Freeland, and Shelley Palen. Without them this wouldn't be on your bookshelf.

Frederick B. Plummer, Jr.

About the Author

Frederick B. Plummer, Jr., graduated from the United States Military Academy at West Point in 1960. He spent his early project engineering days as an officer in the U.S. Army Corps of Engineers, serving in Germany, Vietnam, and as an instructor at West Point after receiving an MS in Civil Engineering from the University of Illinois at Champaign-Urbana.

Following 10 years of service to his country, he returned to Illinois and earned his PhD in Theoretical and Applied Mechanics, while working as a Principle Investigator at the U.S. Army Construction Engineering Research Laboratory in Champaign.

In 1973 he joined Exxon and spent nearly 30 years working as an engineer, supervisor, and manager in the offshore oil and gas business. Nineteen of those years were spent working on large international projects. His assignments included Engineering Manager on the Snorre Field Development Project—the first deepwater platform in the North Sea—and Marine Manager for ExxonMobil Development Company.

After 2 years of consulting and retirement, he joined FloaTEC, LLC, a joint venture, start-up company formed by J. Ray McDermott and Keppel FELS to engineer and build deepwater oil and gas production systems. There he is the Director of Execution Excellence, responsible for coordinating the development of the company's business processes and project management system.

Chapter 1

When Opportunity Knocks

Project engineers are an integrating force in modern industrial society. They link the people who envision the work to the ones who do it. They tackle a job and do what it takes to make plans into reality. Your company or organization consists of departments and groups that specialize in engineering, purchasing, construction, manufacturing, accounting, and other skills. Within those departments and groups are specialists, such as process engineers, machinists, accountants, ironworkers, and more. All add value in their own specialty. Project engineers, on the other hand, have overall responsibility for a certain part of the work or maybe even a small project. They add their value by coordinating and integrating everyone's contribution into an end product. They solve problems through reasoning, teamwork, and leadership. They communicate, and they ask important questions, such as

- Is the work being done correctly?
- Does it cost too much?
- Is someone falling behind?
- Are the working conditions safe?

They lead, and they get the job done.

As a young engineer, a project engineering assignment is a golden opportunity for you. It's a chance to demonstrate that you can accomplish a job through influencing, coordinating, and leading the efforts of others. Most of those who succeed in project management positions started in the trenches as project engineers. This kind of grassroots leadership experience is hard to obtain later in your career and can be a first step toward a supervisory position, if you do well.

1

WHERE DO YOU START?

Early in my career I took over a supervisory position from a guy I will call Jim. He was a bright, articulate veteran and a seasoned supervisor. As we discussed the job, he must have sensed my uneasiness about following in his footsteps. Eventually, he steered the conversation to some fundamentals I will never forget. He said that a person had to do three things to get ahead in the company:

- Find out who your boss is.
- Find out what he or she wants.
- Do it.

To many of you this will seem logical, if not somewhat simplistic, as it did to me at the time. Others may mutter, "That's just kissing up to the boss." No, it's not. "Yes-people" don't always produce results valuable to the organization and aligned with what the management wants. In fact, most effective managers appreciate employees who are strong enough to offer their opinions because their suggestions often lead to improved results. The challenge is knowing how to fulfill the expectations of your boss while you navigate simultaneously through the demands of other stakeholders who are, in a sense, also your bosses. Nowhere is this more the case than in project engineering.

YOUR BOSS(ES)

As a recently hired engineer beginning a new job, you find it easy at first to know who your boss is. Many companies give you a clear assignment and a mentor to get you started. As you develop competence, you're given more responsibility. Eventually, likely you will be handed a task that involves coordinating the efforts of others—a project. You have become a project engineer.

Imagine you are a mechanical engineer who has been working in the Engineering Department of a moderate-sized engineering and construction firm for over six months. Your boss is Bill, the engineering manager, who heads the department. There is a mechanical lead engineer, Walter, who checks all the mechanical work for the department. Walter has been with the company for a long time. He gives you a lot of technical advice and mentoring but doesn't supervise you directly.

Your company has recently been awarded a job to design and build a compressor station for a natural gas pipeline. Sara, in the Projects Department, has been designated as the project manager. Bill has assigned you to the Compressor Station Project, working directly for Sara. You are the project engineer for the compressor package, the key component of the station that consists of the gas compressor, motor, control systems, and the enclosure that contains them.

On the organization chart, your boss is Bill, the engineering manager, who is responsible for supplying people and technical tools (design processes and computer programs) to the projects. The engineering manager rates your performance and manages your promotions and career. In many organizations the engineering manager may be responsible for the technical quality of the engineering work, as he is in this case.

On the other hand, Sara, the project manager, is the one who is accountable to management for getting the job done. She has to be happy with the safety, progress, quality, and cost of the compressor package. Sara is your boss too.

As you may know from experience, this is what looks like an impossible situation. You can't serve two masters. But here you are with two bosses, each with differing requirements, and you must find a way to keep them both happy.

WHAT DO THEY WANT?

Now let's continue by finding out what your boss wants. But wait. You have two bosses, and you have to find out what both of them want. So what's the way forward?

The best thing to do in this situation is to sit down, by yourself, and list the objectives you want to achieve, taking into account what both bosses want. Then review those objectives with each boss, individually. Ask for their input and be willing to accept their comments and work out compromises. It may take a while, but it's time well spent. Above all keep your credibility high and don't lose your cool or your confidence.

As you go through this process, don't forget about Walter, the mechanical lead. While he's not exactly your boss, he is a source of advice and experience, since he has been through this many times before. Having his support will help—especially down the road when you need him as a sounding board to solve problems, or when he checks your work.

Of course, you can't just jump into the middle of writing your objectives. You need information. You have already had a conversation with Bill when he assigned you to the job. He stressed the fact that the client has been having operational problems with the type of compressors that were used on the previous job because of manufacturing quality issues. Bill wants you to make sure that those machines meet or exceed the client's requirements. After that meeting you went to Walter to learn more about the problems on the last job. While meeting with Walter, you took the opportunity to ask him to help you brainstorm your objectives.

Next you met with Sara to introduce yourself and get her thoughts on the project. She took up most of your 10-minute conversation stressing that safety is the highest priority. She also stressed that the client has imposed a large penalty in the contract, in case the project isn't finished by the completion date. (Walter had mentioned that the compressor is "long-lead equipment," so you make a note that the schedule will be important.) Finally she told you that the compressor package must come in within the budget.

Sara plans to have a kickoff meeting with the project team next week. After Sara's kickoff meeting, you will likely have enough information to write a draft of your objectives and start the process of reviewing them with Bill and Sara. Already you can see that you will have a challenge balancing quality, schedule, and cost.

The good news is that you realize this challenge now—not a year from now, when a balancing act you weren't aware you had to do falls apart, and all your bosses are unhappy with you. Forewarned is forearmed.

Keep in mind one more consideration when trying to find out what your bosses want. At times, their requirements may be out of bounds. They may not be aligned with company policy, or they may be asking you to do something unethical or illegal. In such a situation, you will have to know who *you* are. We'll get into this later in Chapter 7 when we talk about business ethics. This should be a rare situation for most engineers. If not, you're working for the wrong company.

DO IT!

With knowledge of who your bosses are, and with an agreed set of objectives, life on the project becomes easier. Now you can turn your attention to getting the job done. That's what the rest of this book is about.

We'll consider the roles and the duties of effective project engineers. We'll focus on the fundamental aspects of getting a job done from the time you start to plan the first activity, until you turn the final product over to the customer.

You'll gain a working knowledge of how to develop your own competence, personal effectiveness, and business judgment while you struggle with all the other demands of the job. We'll explore the concepts of teamwork, leadership, and management. Even though you are a new engineer, such skills are important to your success. We'll even consider the rudiments of office politics, competition with your peers, and an overview of business ethics. A case study illustrates all of those concepts in a relevant, understandable project setting. Chapter 9 offers advice from other project professionals and managers who started their careers as project engineers.

A common thread throughout is the issue of dealing with other people. It's present in the everyday contacts with management, peers, and the people whom you lead. It's present, in spades, in the interactions with people from other cultures, either in your own workplace or in far-off places around the globe. Chapter 8 provides an awareness of the challenges of working internationally and suggests how to deal with culture shock. It also introduces you to a set of cross-cultural communication skills and offers a simple approach to resolve cultural differences.

In all of these topics, the emphasis is the same: to make you aware and offer tangible steps you can take to enhance your job performance. This book is a *toolbox of skills, strategies, and options for development* that you'll find useful —no, essential—for launching your career. It's not every tool you'll ever need, but it's something you can carry with you while you're getting started and probably for the rest of your life.

Now let's consider what project engineers do in a little more detail, but still at a fundamental level.

Chapter 2

What Do Project Engineers Do?

TOTAL AREA RESPONSIBILITY

As a project engineer, your overall role is rather straightforward. You are totally responsible for everything that has to do with your area!

The idea of total area responsibility is more of a mind-set than anything else. It's something you have to become comfortable with, without being arrogant. If you do it well, you'll be recognized and respected by the rest of the project—and others outside the project—as the go-to person for your area.

Total area responsibility includes planning the work and controlling it. It involves handling whatever comes along. Sometimes there are disturbances to quell, or disagreements to mediate, or problems to solve. There's an aspect of doing whatever it takes, but priorities have to be set to stay within deadlines, budgets, and the realm of what's humanly possible. I was on a project once that had as one of its ten principles, "Good enough is usually the best." That's excellent advice when one considers most projects entail an avalanche of work, and each one of us has a personal life to live. Prioritize, or you'll be buried by the flow!

TYPES OF AREAS

To get a grasp of what an area is, let's consider a few examples.

Specific Part of a Facility, or a System

For a design or construction project engineer (on either the client's or contractor's team) an area is often a specific part of a facility—for example, a part of a building, the engine room, the control room, the living quarters, a structure, and more. The area could also be a system, like the process system, the electrical system, or several of the instrument and control systems.

Main System or Assembly

In equipment manufacturing companies, the project engineer's area could be a package, which is either a main system or a main assembly of parts for the product. The project engineer or package engineer then becomes the steward for the package as it passes through the design, development, manufacturing, and acceptance process. The package engineer doesn't do the design, or the manufacturing, or the testing. He or she leads a team that develops the design requirements and then ensures that the design drawings and specifications are produced in accordance with those requirements and the project schedule. The package engineer also works the interfaces and may be responsible that the quality of the end product satisfies the requirements and the client's interests.

Purchase Order

On the client's or contractor's team, a project engineer can also be responsible for administering purchase orders for either equipment (gas turbines, generators, pumps) or bulk materials (structural steel, piping materials, wire, control valves). Here the area is everything that is delivered under the purchase order. This type of project engineer is also called a package engineer, but has a somewhat different job than the manufacturer's package engineer. The purchase order package engineer has more of an oversight role in the quality surveillance and expediting process. He or she ensures that equipment is satisfactorily tested at the factory and delivered on time. The package engineer is commonly paired with a purchase order administrator (buyer) from the Procurement Department, to achieve the proper balance between technical and commercial skills.

DEFINING THE AREA

Some managers are careful about defining the areas of responsibility for their project engineers so that nothing falls in the cracks between them. Others are not. In either case, however, it's left to the project engineer to work out the details. Remember...total area responsibility.

One of the keys to successfully defining an area is to ask questions. There are always questions concerning the scope of the work. Is the job only engineering, or does it include engineering and procurement? Is it only construction, or are there responsibilities for preparing the fabrication drawings? What are the responsibilities when the package moves into production (manufacturing)? Will the project engineer supervise the acceptance testing? Are there commissioning and start-up responsibilities? The project engineer should keep asking questions until the scope of responsibilities becomes clear.

He or she will also find it useful to define the boundaries of the area, especially at the interfaces with other parts of the project or other parties outside the project. A project engineer can't afford time- and energy-consuming confusion at the boundaries of an area. Meeting frequently with the interfacing parties, asking questions, resolving issues, and documenting agreements are all part of defining the boundaries. Even within well-defined, nonproblematic boundaries, the project engineer's duties are complex and challenging.

THE PROJECT ENGINEER'S DUTIES

As you've probably already realized, the nature of the project engineer's overall role is significantly influenced by whether she or he is working for the company that is doing the work (contractor and manufacturer) or the company that is buying the work product, and thus overseeing the work (client):

- Contractors' and manufacturers' project engineers are primarily responsible for the work on a specific part of the job.
- Client project engineers are in more of a watchdog role over a large area.

A project to build a cargo ship can serve as an example. The shipyard will have hundreds of engineers working on the job. A shipyard project engineer could be, for example, responsible for the engine room and would plan and control all the engineering in that area. The client would likely have a small team in the shipyard. One of them would probably be a Hull Project Engineer responsible for overseeing the design and construction of the entire hull. *Despite the differences in roles, the duties of all project engineers are similar.*

Maybe that's why there's a natural tendency for the clients' project engineers to encroach on the responsibilities of the contractors' project engineers. It's difficult for contractors and manufacturers to push back in those situations without appearing to lack a customer orientation. If they don't, however, the split of responsibilities between their project engineers and the client's will remain blurred, resulting in wasteful duplication of effort.

Let's work together through a set of project engineers' duties, keeping in mind that there's no unique way to be a successful project engineer.

Every job is different, and every project engineer will likely handle a given job differently. When people and politics are added to the equation, it gets even more complex. Nevertheless, it's important to have a list of duties to remind you to *concentrate on all the facets of your job.* Without the list, you'll be drawn to the crisis of the moment, and you will forget the balanced approach that leads to achieving excellent results. So take this list of duties as your starting point, and create your own as you approach your project engineering job:

Duties of a Project Engineer

- Plan and control the basic work.
- Lead safety.
- Identify, assess, and mitigate risks.
- Achieve quality standards.
- Control schedule within the plan.
- Control costs within the budget.
- Control interfaces.
- Manage changes.
- Solve problems and commercial issues.
- Lead the effort.

Remember, your role is to exercise total area responsibility! If one of your specific duties isn't on the list, add it.

Let's look at those duties in more detail. Some of the duties that may not be familiar to most new engineers are covered in a little more depth.

PLAN AND CONTROL THE BASIC WORK

The fundamental responsibility is to get the basic work within your area *done correctly the first time.* For the most part, other people will actually do the work. The project engineer's duty will be to coordinate and control all of the activities, so that they happen safely, on time, and within budget.

There are both technical and commercial dimensions to the job:

- Technical considerations to ensure that the engineering, manufacturing, or construction is carried out correctly
- Commercial tasks such as administration of the subcontracts and purchase orders to ensure on-time delivery of a quality product, and to prevent excessive charges and claims for more money from suppliers

Generally, it all boils down to planning the job, and then actively following up as the work progresses.

LEAD SAFETY

Safety is arguably the most important aspect of modern industrial life. Excellent companies relentlessly pursue the goal of an accident- and illness-free workplace. When they succeed, they raise the bar. Safety becomes a fundamental part of the company's culture.

The project engineer is in a position to be a safety leader. Safety in the design is crucial to the long-term safe operations of the facilities. The project engineer can work closely with discipline engineers, construction engineers, and operations people on the design team to identify and mitigate hazards through design reviews and HAZOP exercises. (HAZOP is a hazard and operability study to improve the safety aspects of a design.)

Safe behaviors during construction, commissioning, and start-up are essential for the prevention of fatalities and lost-time accidents. During those phases, the project engineer can stress the importance of effective job safety analyses (JSA) by employees, since their involvement in safety is crucial. When walking around the job site, the project engineer can discuss safety with the workers and set an example by using the required personal protective equipment for that area (hard hat, safety glasses, and whatever else is required).

And how will you know when you understand safety? You'll have developed a genuine concern for the safety of others and yourself. It will be your first priority, consistently. You won't get so wrapped up in looming deadlines and everything else that you lose sight of safety considerations. You won't dart out into the site without a hard hat "just once" because you're in a hurry. And if it becomes necessary, though awkward, to speak up about an unsafe design or unsafe work practices, you will.

IDENTIFY, ASSESS, AND MITIGATE RISKS

The best project engineers and managers constantly manage the risks around them. When they plan an approach, they think of a second or third way to accomplish the same objective if the first fails. They ask, "What if?" then think through their contingency plans. When adverse events occur, they instinctively perceive the magnitude of the threat and imagine ways to reduce or eliminate undesirable consequences. Sometimes they realize a new opportunity in the midst of chaos. Stuff happens, and when it does, the best project engineers anticipate risks and respond effectively.

Risk management techniques help project engineers perform well in a variety of situations:

- Manage safety, health, and environmental issues or incidents
- Apply new technology

- Evaluate contracting and purchasing strategies
- Assess the risks of bidding opportunities
- Ensure safety in designs
- Manage changes to designs
- Address project implementation risks in harsh environments like the Arctic or the remote regions of West Africa or Southeast Asia
- Evaluate inherently hazardous construction operations
- Prepare for the start-up and operation of plants, other facilities, and vessels

The JSA (job safety analysis) mentioned in the safety section is an excellent example of risk management at the working level. Effective construction project engineers encourage construction supervisors or task leaders to get their crews together before starting the day's job. They talk through each step of their daily task to identify hazards and ways to avoid those hazards and work safely.

Risk management is not easy and often not fun, but it's critically important. Nobody and no organization wants to end up in front of an accident investigation board explaining what they did wrong or failed to do right. So consider three aspects of risk management that you can use as a project engineer from day one:

- Risk identification
- Risk assessment
- Risk mitigation

Those are the fundamentals of a qualitative risk management approach that will be useful to you on a day-to-day basis. Let's unpack these.

Risk Identification

Early in any project, large or small, it's good practice to hold a structured hazard identification (HAZID) exercise. The Project Manager or designated representative calls together a group with broad experience in the project's critical areas. They brainstorm events or incidents that can create major adverse effects on SH&E (safety, health, and environment), quality, cost, schedule, and even their organization's reputation. They generally meet in group sessions that stress openness and ensure that no one's ideas are rejected or judged inferior. The output is a Hazard Register that includes preliminary assessments of the consequences and probability of occurrence for each hazard. Often the group will prioritize the list so that hazards with the greatest risk are at the top of the list. They may even establish a short list of risk issues that deserves senior management's attention and stewardship.

This HAZID exercise will get the team focused on the risks and at the same time build commitment toward addressing them. When everyone's ideas are respected and considered, teamwork is a valuable by-product.

The HAZID exercise easily scales down to the project engineer's area. At the beginning of the job, and at the start of each major task, the project engineer can call a brainstorming meeting to

- identify the risks,
- estimate probabilities and consequences for each (on a high-medium-low basis), and
- rank the list from worst to least.

Those HAZID sessions will have the same benefits of focus, commitment, and teamwork as at the project level.

Risk Assessment

There are two basic types of risk assessment—qualitative and quantitative. The qualitative approach is the most widely used.

Quantitative risk analysis (QRA) can be equally valuable, but it's often a matter of organizational policy as to when or whether it's used. The quantitative approach requires the organization to have policies and criteria against which to assess risks. For example, the criteria might specify that engineers must design for any event with a probability greater than 10^{-4}. If a certain type of electrical short has a calculated probability of 8×10^{-3}, the design of that circuit must be improved to avoid the short or move its probability below the criteria. Quantitative risk assessments generally require experts to do the analysis, using comprehensive databases compiled from years of industry operations. If your organization uses quantitative risk analysis, you can read the internal documentation to learn how QRA is done and what your role is.

Qualitative risk assessment is more subjective, but still addresses the core issue of which risks are acceptable and which are not. The qualitative approach engages key people across the whole organization and brings their judgment and insights to bear.

One popular tool for assessing risk is the risk matrix (see Figure 2.1). The risk matrix can be used with either qualitative or quantitative risk assessments. It's a handy way of portraying the risk of several events by plotting the probability of occurrence versus the severity of the consequences, as shown in Figure 2.1. This particular matrix is arranged so that the worst events are in the upper-left corner, but that is strictly preference. By using five categories on each axis, it's possible to separate events into three zones of risk:

- Black—disastrous or catastrophic risks that must be avoided or mitigated to a lower zone
- Gray—moderate risks that should be mitigated by design improvements, control measures, or operational procedures to the extent possible
- White—lower risks that can certainly be further mitigated but are likely acceptable as they are

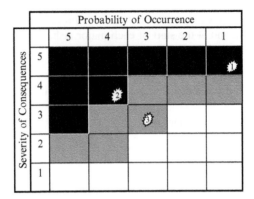

Figure 2.1   Risk matrix.

The risk assessment team may use the Hazard Register as a starting point, but will probably dig much deeper to make sure they have captured all the risks. For instance, they might work step-by-step through systems (e.g., high-voltage power system), design areas, production processes, construction sequences, bid invitations, start-up procedures, or event scenarios (e.g., storms, floods), trying to find any disaster that could befall them or the client. Mature organizations will have checklists, procedures, or best practices to assist their risk assessment teams. Nevertheless, the team should always look outside the box.

The assessment team will estimate the probability that each event will occur using their collective judgment and consensus. Risks that are almost certain will be placed in the left column (5), while risks that have a nearly negligible probability fall in column 1 on the right of the matrix. Risks that are unusual but possible are placed in the center column (3).

The team also assesses the severity of each event. The five rows of the risk matrix indicate the severity of consequences, with Category 5 being the most disastrous and 1 the least. As indicated in Table 2.1, severity depends on what is being assessed. For projects, Category 5 events have consequences that are a crisis, while Category 1 events have negligible consequences. Category 3 would represent a moderate but significant loss. You will have to provide your own criteria for severity when doing a risk assessment, since every situation is different, but this table for a $500 million project should add to your frame of reference. Other factors such as chemical spills, damage to reputation, product recalls, legal exposure, operational downtime, and more can be evaluated on this same scale.

Once the risks are assessed, the most significant ones are plotted on the risk matrix. Three numbered examples are shown in Figure 2.1. The next step is to eliminate or mitigate them.

Table 2.1

Severity of Consequences

Category	Injuries	Cost	Schedule
5 Crisis	Many fatalities	$5 million	3 months
4	One fatality	$1 million	2 months
3 Significant	Lost-time accident	$250,000	1 month
2	Medical emergency	$100,000	2 weeks
1 Negligible	First-aid case	$20,000	1 day

Risk Mitigation

Risks can be reduced by three general types of measures:

- Avoiding the event altogether
- Reducing the consequences to an acceptable level
- Decreasing the probability it will happen

With ingenuity, it's usually possible to find a better solution, but it may cost more or take longer to construct. In some cases (not involving safety risks), insurance could be a better alternative than extensive design changes. Let's consider a few examples of how to mitigate risks from the design risk assessment for a factory project.

Event 1

An observant engineer on the risk assessment team noticed a lack of redundancy in the roof structure of the plant. Further investigation indicated that during the 100-year hurricane the roof would collapse, essentially destroying the factory. Even though this event is unlikely, the consequences are so severe that the design must be improved. By strengthening the roof truss, Event 1 would move down on the risk matrix well into the white zone, while the probability would remain the same. This would prevent an escalating event in which one failure leads to a worsening disaster.

Event 2

An operations supervisor on the team spotted a large, manually operated valve located beside a catwalk directly above the assembly line. A short operator would find it awkward to open or close the valve, increasing the risk of a fatal fall. Normal maintenance or replacement of the valve in service would require shutting down the assembly line and building a scaffold. Event 2 was moved into the white zone by relocating the valve to facilitate its operation and installing a lifting device to maintain, remove, and replace it.

Event 3

Pipes operating at 200°F had been routed along a walkway that runs the length of the plant floor. The assessment team recommended that those pipes be insulated or caged in to prevent personnel from being burned. They could have recommended control procedures such as warning signs, but passive means of assuring safety are usually preferable to active measures.

Hopefully by now your sense of your risk management duty is beginning to take shape.

Project Engineers' Risk Management Role

Your company or organization probably has a Risk Management Procedure, and most good projects have a Risk Management Plan. Project engineers should familiarize themselves with those procedures and plans because they play a vital role in implementing risk management within their areas as the project unfolds. If you're in a small company that doesn't have those documents, you will have to find a good book on risk management or hire an expert.

Project engineers proactively manage the risks for their areas using some variation of the process explained above. For an area where hazards are present or safety is paramount, risk management is a crucial part of the project engineer's job. Take, for example, the design of high-voltage power systems, nuclear facilities, or processes that handle flammable or poisonous agents. There it may be prudent or even mandatory to carry out independent safety studies or quantitative risk assessments to demonstrate to management, the public, and most importantly yourself that the design is safe.

Project engineers form the core of the talent convened at various stages of the project to conduct the formal risk assessments that are called for in the Risk Management Plan.

In the early planning phases of the project, the project management team emphasizes finding and eliminating hazards that threaten commercial or economic viability. Selecting the wrong technology could spawn a research project while the project team is trying to build the job or produce a new product. Or choosing an infeasible concept can stop the project dead in its tracks when the truth becomes known. Obviously, the earlier risks are detected and mitigated, the better.

But it's not enough for project engineers to identify hazards, assess risks, and mitigate them. They must proactively ensure that their area has the requisite quality built in.

ACHIEVE QUALITY STANDARDS

The way that organizations approach quality varies widely. Project engineers must gauge this and perform accordingly. At the high end of the quality scale you will find manufacturing companies. Motorola and GE are among the firms that have made the Six Sigma Quality Approach popular. That approach seeks to reduce process variations to 3.4 defects per million by concentrating on

- process improvement,
- process design or redesign, and
- process management.

The goal is to eliminate the root causes of defects, or to control the processes so that defects don't occur.

For some organizations such as those in the nuclear industry, quality is an essential part of safety and receives enormous management and governmental attention. They take a similar process-oriented approach to eliminate all defects.

In all those high-quality environments, it's crucial for engineers to virtually become quality experts. That's really beyond the scope of this book. Instead I'll offer some concepts and basic tools to help new project engineers achieve the required quality standards for their areas of a project. For the most part, companies and organizations adhere to the quality management system requirements described by the International Organization for Standardization in their standard ISO 9001.

Quality Management

Quality starts with management. Management provides the human and material resources and then directs the planning, designing, and implementation of the production processes. The people who do the work must be trained and motivated to produce a quality product—to build in quality. The production of the product is carefully controlled to ensure that end products meet requirements, including the client's expectations. The final piece of the quality cycle is a program to measure the work processes and products, analyze the results, and constantly improve.

Most companies have a quality management system in place. An engineering company's formal design management procedure for preparing, checking, and approving drawings is an example of one procedure in a management system. Part of the project engineer's quality role is to make sure that the quality systems are functioning for the work that's done in her or his area. The project's Quality Manager or one of the quality engineers will be happy to explain the quality processes and acquaint you with the project's Quality Plan.

The Quality Manager often reports directly to the Project Manager. This gives the quality group more authority and independence as they go about assessing or auditing the project's systems. When a quality audit team finds a serious fault in a project system, they issue a *corrective action*. The project team is then required to improve the process in question and report that the corrective action has been resolved. Project engineers should follow up to make sure corrective actions that apply to their areas are implemented.

Design Quality

On projects, the foundation of quality is the design. Problems in the design will eventually cascade into procurement, construction, manufacturing, commissioning, or operations. Eight internationally recognized design quality steps (ISO 2000) are used on projects and in industries around the world:

- Planning and control of the design process
- Preparing design inputs
- Reviewing design inputs for accuracy and omissions
- Preparing design outputs
- Performing design reviews
- Performing design verification (confirmation that requirements are met)
- Performing design validation (overall demonstration of the design, such as factory acceptance test, commissioning, performance testing)
- Managing changes to the design

With the exception of the first bullet, project engineers will contribute to most of those steps. The steps are usually prescribed in one way or another in the project's Quality Plan.

(Copies of ISO 9001 and all ISO standards can be purchased from the American National Standards Institute, 25 West 43rd Street, NY 10036; phone: 212-642-4900; e-mail: info@ansi.org; webstore: http://webstore.ansi.org/ansidocstore/default.asp.)

Preparing and Reviewing Design Inputs

One of the project engineer's most important quality activities takes place at the beginning of the job. Project engineers must understand the contract and review the client's design inputs that apply to their area. In addition, they often coordinate the compilation of the internal requirements needed to design the product or facility. This is usually part of an overall review led by the Lead Project Engineer or Engineering Manager. It consists of verifying that the functional requirements, design basis, specifications, and conceptual design are correct and reasonable to achieve. Appropriate standards, codes, and regulations are identified. Sometimes clients don't give clear and complete requirements. If problems or mistakes are uncovered, project engineers have an obligation to point them out and process them through the design change process.

Preparing Design Outputs

During the preparation of design drawings, data, and documents, the real quality is built in. Companies implement quality management systems to assure that the designers function as a team and coordinate details between themselves, but it's mostly the skill, diligence, and competence of the design engineers that produce quality in the product.

Performing Design Reviews, Verification, and Validation

These three quality steps are a safety net for the project. When the design is nearly complete, a *design review* is called, which assembles key design engineers, project engineers, supervisors, managers, and more. The group works through the design in a structured way looking for issues. It's a best practice to have participants submit written questions ahead of time so that discipline engineers can be prepared to address them during the design review. *Verification* consists of planned activities to confirm that critical design outputs meet their requirements. When reviewing the design input requirements, project engineers will compile a list of critical detailed design products that must be verified later in the design process, but before manufacturing or construction starts. Verification can be anything from independent calculations to model tests. *Validation* is an overall test that demonstrates that the design functions as intended. An example would be a factory acceptance test for an electric generator that runs the equipment under simulated operating conditions.

Managing Changes to the Design

This quality step is so crucial that it will be considered later as one of a project engineer's primary duties.

Measurement, Analysis, and Improvement

Measurement, analysis, and improvement are critical parts of the quality management systems for organizations in the manufacturing, fabrication, or construction businesses. For example, project engineers working for construction contractors coordinate the activities prescribed in the Inspection and Test Plans (ITPs) for their areas. An example is taking precise measurements (dimensional control) to confirm that the dimensions are accurate for parts that must fit together. Again, project engineers don't make the measurements. They make sure that the required quality control processes, prescribed in the ITPs, specifications, or other project documents, are implemented and functioning.

When a work product has a serious deficiency, the system will issue a *noncon-formance*. A nonconformance is a big deal. Project engineers should chase any nonconformances that affect their area. In this situation, project engineers have the main responsibility for coordinating the corrective measures and pressing all the parties to get the corrections implemented in a timely manner.

Surveillance

Surveillance describes the quality management roles for a client's or contractor's project engineers when their organizations are in an oversight role. Surveillance consists of following up to ensure that the contractors' or vendors' quality management systems are in place and fit-for-purpose. It's done by spot-checking test results to confirm that requirements are being met and that corrective actions and nonconformances are being fixed.

Part of this surveillance is informal monitoring, which takes place while the project engineers are walking around and asking questions in the office, on the shop floor, or at the job site. But the more important activities should be planned. One item in the project engineer's Surveillance Plan could be for experts to review the welding procedure qualification documentation before welding starts. Other vital planned surveillance might include technical audits, design reviews, inspections (using detailed checklists), or witnessing factory acceptance tests. These checks should be done when an engineering, purchasing, manufacturing, or construction work product is nearing completion but before that work product is used. For example, a piping expert could check the critical, high-pressure piping drawings as they are being completed and before they are sent to the shop or fabrication yard.

Prioritizing

As I mentioned at the beginning of this chapter, project engineers won't be able to check everything. That's impossible and doesn't make economic sense. Project engineers only have time to decide what's critical for the safety, operations, and economy of their areas and then check those priority items. Too much checking increases cost and extends delivery.

CONTROL SCHEDULE AND COST

Monitoring and controlling schedule and cost are the two project engineering duties that receive the most management attention. Volumes are written about the tools and techniques to manage schedule and cost. However, your company or organization will have practices, procedures, and probably proprietary cost and schedule software that have grown up over time. These are tailored to the business, and you will have little choice but to use them. With that in mind, let's start with some concepts to familiarize you with the cost and schedule landscape and give you a few hints. The case study in Chapter 5 illustrates how project engineers manage cost and schedule in practice.

Project engineers are on the front lines in the battle to maintain progress and stay within the budget. In an ideal world, project engineers would like to start with cost and schedule estimates that they are confident of achieving. The estimates probably won't be "bulletproof," but they'd better have their helmet and flakjacket on. Surprises happen, and some reserves are needed to deal with them, without busting the budget or delivering late.

Control (or baseline) schedule and cost estimates are generally established at the beginning of detailed engineering. The sequence is

1. Planning
2. Scheduling
3. Budgeting

By that time in the project, plans are usually in place, and the scope and sequence of the work known.

Scheduling

When the plans are in place, it's time for project and discipline engineers to break the work into detailed activities, and then estimate the personnel resources, durations, and costs for each activity. "Float" or extra time should be included in the durations to give flexibility for solving problems. Engineers are notorious for underestimating how long activities take to complete. It's not surprising for tasks to take twice as long as engineers plan. At your level it's wise to underpromise and overachieve.

Engineers must also coordinate with others to anticipate

- which activities must be completed before starting a given activity, and
- which can't be started until that activity is finished.

For example, the project shouldn't start manufacturing engine blocks before their design drawings are received by the shop. On real projects this sort of thing happens, but it shouldn't be the base plan that is included in the control schedule. Risks are introduced when logical dependences between activities aren't observed. Fast tracking—as this is called—leads to trouble when the people actually doing the work don't have design information, equipment, or materials when they need them.

Next, a planner-scheduler aggregates all the activities into a schedule that goes through several cycles of peer and management reviews. The project management will likely add some schedule reserve, at the end of the schedule, to cover unforeseen circumstances. For the project as a whole, the schedule estimate, including the schedule reserve, should be a fifty-fifty estimate, which means there's a 0.5 probability that the schedule will be achieved. If there is no reserve at the end of the schedule, it's probably a target estimate, with no margin for error. In some critical circumstances higher confidence, say, 90% probability of success, makes sense.

The project establishes milestones at important events, such as "start of construction" or "factory acceptance testing." Project engineers should set intermediate milestones for their areas to help them achieve the project milestones. Actual progress is stewarded against these intermediate milestones and reported monthly along with forecasted completion dates. The client may negotiate incentives and penalties at key milestones, but even if there are no penalties or incentives, missing a project milestone is a serious matter and an important warning sign.

Cost Estimating

Once the schedule begins to take shape, budgeting begins. Cost engineers consolidate the engineering, procurement, and construction estimates for each activity into the control (baseline) estimate. Allowances are added for specific activities that are less defined and subject to cost growth. Allowances bring those activities' estimated costs up to the level where the responsible engineers believe they will be at the end of the job. For tasks where there is a range of cost uncertainty, it's wise to choose the larger sum at this stage of the game.

As with the schedule, the overall project cost estimate should usually be fifty-fifty, with an equal chance of over- or underrunning the budget. To achieve that, project management adds contingency for the unknown things that can happen, but the project engineer won't be able to use any of that contingency without management's approval. When contractors prepare bid estimates, they do a detailed risk assessment and then add risk money into their bid to cover the impact of known risks.

As the job progresses, the project engineers steward the actual costs incurred for their areas against the planned budget. Planned and actual costs, along with a forecast of the cost to complete the job, are normally reported on a monthly basis, if not more frequently.

Benchmarking and Risk

Good projects benchmark their cost and schedule estimates against previous project data, and then adjust cost contingencies, allowances, and schedule reserves to bring the overall estimates to the desired level of confidence. Suppliers (contractors and vendors) facing penalties would want a high certainty of success—perhaps 90%.

Suppliers are under tremendous pressures during the early phases of a project while they are bidding. Technical definition at this stage is typically incomplete. Competitive forces make them bid low, but if they substantially underestimate costs and schedules in their bids, they could face huge losses on the job. Benchmarking costs and schedules against past performance is one way to mitigate those risks.

Contingency for unknown occurrences is added to the base estimate to bring the estimate to the benchmarked level. In addition, special allowances are often added to the benchmarked cost to cover known risks—on a risk-weighted basis. Risk weighting is used because project management teams realize that all risks probably won't happen. They therefore estimate the cost impact for each known risk event and multiply it by the probability that risk event will occur. This gives a risk-weighted cost for each event. These are summed and added to the cost estimate as a special allowance or "risk money."

Benchmarking and contingency are the suppliers' first lines of defense on the cost front. Contractual terms, commercial strategies, and risk management may also help, but unforeseen events can have severe consequences. Project work is always a balancing act.

BALANCE THE SAFETY, QUALITY, COST, AND SCHEDULE PRIORITIES

A balance must be struck between the four important project objectives—safety, quality, cost, and schedule. There is a widely held view that you can't achieve all of these at the same time. You have to prioritize. Safety is, without question, first priority. It is fundamental and can't be compromised. Given that, you can probably manage to have two of the remaining three:

- Cheap and fast, but at low quality
- A quality product fast, but not cheap
- A quality product cheap, but not fast

Most would probably agree that the first possibility—a low-quality product that doesn't meet the requirements—is not acceptable. In fact, it could be in breach of the contract. But there are situations where quick and cheap can make sense. I was reminded by an experienced construction site manager that temporary structures or systems that are used only for a short time during construction need only be safe, cheap, and quick. Nevertheless, the project engineer should conduct a deliberate evaluation involving credible experts before compromising safety or quality for cost or schedule. *For the most part, the project engineer is left with trading off the priorities of cost versus schedule.*

Before leaving this subject, I would like to introduce the concept of "fit-for-service," since it is the ultimate balance of quality, cost, and schedule. Sometimes, especially near the end of a large construction or manufacturing job, a serious quality problem is discovered. For example, several hundred weld defects could be found that would take millions of dollars and several weeks or months to repair. In this situation it is generally best to call in credible experts to do a fit-for-service evaluation before deciding to carry out the repairs. Even though the weld defects don't meet the specifications, it may be possible to prove that at least some of the welds are adequate for the specified design conditions. Most clients would rather accept a well-documented, fit-for-service evaluation than miss a project milestone.

This discussion of balancing safety, quality, cost, and schedule is certainly not a rigorous argument. Take it for what it's worth. Unfortunately, for some strange reason, most executives and project managers haven't heard of this idea of trade-offs. They want it all—high quality, low cost, and world-class schedule. So...good luck!

CONTROL INTERFACES

Many experienced engineering and construction managers believe that controlling interfaces is the biggest challenge they face.

At an interface there is another party on the other side with different drivers. That other party is also probably working toward a different schedule. For example, the first piece of information you need from that party could be on the last drawing they are scheduled to produce.

The Nature of Interfaces

Here is an example of an interface from the offshore oil and gas industry. Deepwater, floating oil, and gas platforms often have subsea wells tied into them, even one like the Tension Leg Platform in Figure 2.2, which supports surface wells. These subsea wells are drilled thousands of meters into the seabed. The wellheads, which contain the wells' pressure, lie at the mud line hundreds of meters below the ocean's surface. Flowlines and risers connect the wellheads to the floating platform, which holds the oil and gas processing equipment. There is an important interface between the subsea contractor and the platform contractor at the point where the riser connects to the platform.

The subsea contractor begins the design at the bottom of the drilled hole by sizing the well and designing its detailed mechanical components. Next they design the wellhead, flowline, and riser. At that point they can design the connection to the platform and figure out the details of how the risers will be installed.

On the other side of the interface, the platform contractor immediately needs details on the riser connection and the loads it will impose on the floating platform,

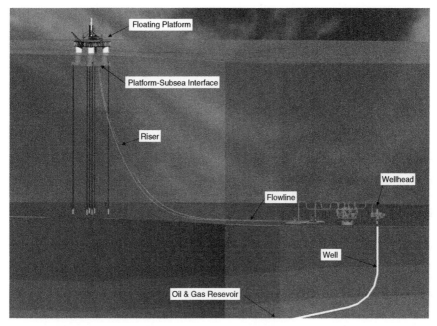

Figure 2.2 Offshore floating platform with subsea oil and gas well.

since its hull is one of the first items ordered. Of course, those design details aren't available and aren't scheduled to be done by the subsea contractor for months. The mechanical and layout discipline wants to know what equipment must be installed on the platform deck to service the wells and install the risers. The process and piping engineers need to know the flow parameters (e.g., temperature, pressure, flow rate) crossing the boundary. Electrical engineers must have the subsea power loads. The instrument discipline needs the details of the control signals that interface with the platform control system. The list goes on.

Now you can begin to see the dilemma that projects encounter at interfaces. The last component the subsea contractor can design is the first component for which the platform contractor needs the design input.

A Useful Interface Process

The key to successfully handling interfaces is to engage that other party early and develop interface plans and agreements. Here is a simple, three-step process that works as long as there is cooperation between the interfacing parties:

1. Define the interface and develop a plan for drawings and documents that must be exchanged and agreed on.
2. Identify issues (mismatches, clashes, lack of definition, schedule problems).
3. Agree, *in writing,* to
   • the tasks needed to resolve the issues,
   • responsibilities for the various tasks, and
   • a date for completing each task or providing information to the other party.

In our example, the subsea contractor will have to make conservative estimates that must be confirmed as the design progresses. Perhaps some of the process and installation work can be accelerated to give the platform contractor critical interface information.

Early engagement is essential. After contracts and purchase orders are awarded, the problems at interfaces have both technical and commercial implications. These commercial impacts grow with time and can seriously complicate problem solving, since most people aren't very cooperative if they are losing money. When cooperation breaks down, it's time to escalate the problem to management. In Chapter 5, we revisit this important topic of controlling interfaces in the context of a case study.

Interface management and the design process itself inevitably lead to changes that must be considered in a deliberate, controlled way.

MANAGE CHANGES

Another major part of the project engineer's job is the challenge of managing changes that are proposed. Changes can come from either inside or outside of your area. Once the project enters the implementation phase (generally detailed engineering), it is wise to limit changes. On the other hand, if there are serious mistakes in the design, they must be detected and corrected as early as possible. Don't hold to a design that is faulty, just for the sake of not changing anything. The standard usually applied to a design that someone is trying to change is this:

• Is the original design safe?
• Does it work?
• Does it comply with the laws and regulations?

If you can answer yes to these questions, the change proposal should probably be rejected. But, of course, the project management will have to apply the final judgment. It must be kept in mind that it is virtually impossible to anticipate all the adverse effects of a design change, especially once you are into manufacturing or construction. Let's drop in a few months later on our fictitious story about the subsea-to-platform interface situation to illustrate the effects of a change.

The subsea contractor estimated the riser interface loads and the size of the connectors. Both contractors continued their designs based on those assumptions. After the platform contractor had started construction on the hull, the bids came in for the risers. A new supplier had offered a $500,000 savings on the 10 risers. Since the cheaper risers were considerably larger and heavier than those in the design basis, the Project Manager demanded that a design change be processed. Engineering estimated that the changes to the hull would cost $300,000, a net savings of $200,000. The change was approved; however, no one contacted the construction yard to confirm the estimate. Engineering drawings for the changes to the hull and the riser connection porches were engineered and sent to the yard, but not before assembly of the hull had started. Scaffolds had to be built to access some of the changed areas and welding caused considerable paint damage. As they got into the work, field-run cables and piping had to be rerouted, and other changes were needed. The hull fabricator submitted a contractual change order for $1.8 million, which was passed on to the client—a net loss of $1.3 million on the so-called cost-saving change.

It's worth saying again that *it is virtually impossible to anticipate all the adverse effects of a design change, especially once you are into manufacturing or construction.* Your project will have a change management system to control how changes are requested, evaluated, approved, and implemented. It is crucial to follow that system and keep all of the affected parties informed. And be skeptical of late changes because they lead to problems. It's hard to imagine a client accepting an extra cost of $1.3 million, even if their decision caused it.

SOLVE PROBLEMS AND COMMERCIAL ISSUES

Project engineers are the chief problem solvers for their areas, and there's no way around that. Inevitably, problems will occur. They will spring up from within the area as the work progresses. They will come hurtling over the boundary of the area, either at the interfaces or due to changes imposed by the project management or the client.

Most engineers think of problems in technical terms. When the design flaw is corrected on the drawings, or when the manufacturing problem is identified and fixed, the problem is solved. But nearly every problem has a commercial side, too.

Changes to the work and solutions to problems (which are also changes to the work) disrupt what was agreed on when the contract or purchase order was signed. Client teams tend to think that most changes are included in the price that was agreed on when the contract was signed. Some contractors and vendors look at changes as a commercial opportunity to get more money. The truth generally lies somewhere in between. Project engineers soon realize that there are tremendous pressures on both the client's and supplier's teams to control the damage caused by changes or maximize the benefits to their organization.

The primary responsibility for commercial matters lies with the Project Manager and his or her delegated contract representatives. Nevertheless, the project engineer has a duty to solve the problems that affect his or her area in a way that satisfactorily resolves both the technical and the commercial issues. Here is a three-step, seven-substep formula for success:

1. Don't avoid the problem. Problems on large projects have a way of leading to other problems if they aren't dealt with when they become known.
2. Focus on the problem and, if necessary, get experts involved. These experts should be
   - people who have essential information or expertise to contribute, and
   - people who will need to be committed to the solution, when it's time to implement it.

   If possible, try to limit the people involved to these two categories. The problem-solving group should include a mix of technical and commercial expertise, along with representation from major stakeholders in the outcome.
3. Resolve the problem and document the solution. This usually takes the form of one or more of the following:
   - A formal change (within the project's management of change process)
   - A change to design drawings and documents
   - A change to one or more contracts
   - A change to purchase orders involved
   - Possibly some other kind of documentation, such as an interface agreement, which will eventually work its way into the project drawings and documents

Those three steps may seem simple, but they deserve emphasis. Too many times project people turn their backs on problems and fail to take the steps to face them head-on. They would rather put the change orders in a drawer and procrastinate, hoping problems will go away. They seldom do.

LEAD THE EFFORT

Leading is influencing others to accomplish goals in a given situation (Hersey and Blanchard, 1982). Project engineers have ample opportunity to lead. For the most part, this is how they get their job done. Those new to project engineering may fixate on the urgency of the task at hand and forget that the people and situation must be factored in. Effective leaders balance the accomplishment of the task with their relationship to the people involved because both are important. Often, project engineers don't have the organizational authority to fully accomplish their teams' goals. They must earn and maintain the respect, loyalty, and support of the team to achieve success over the long haul.

I have always liked the distinction that Hersey and Blanchard (1982) draw between leadership and management. They see leadership as a broader concept than management, and management as a special form of leadership where organizational goals are paramount. They define management as "working with and through individuals and groups to accomplish organizational goals." In other words, "The achievement of organizational objectives through leadership is management" (Hersey and Blanchard, 1982).

Project engineers often find themselves managing, so they must be prepared. The next chapter is a crash course in management to help you on your way.

REFERENCES

Hersey, P., and Blanchard, K. H., *Management of Organizational Behavior: Utilizing Human Resources* (4th ed.) (Prentice Hall, Inc., New Jersey, 1982), pp. 3, 83.

International Organization for Standardization, *ISO 9001 Quality Management Systems—Requirements* (3rd ed.) (ISO, Switzerland, 2000).

Chapter 3

A Crash Course in Management

A project engineer depends on the work of others to get the job done. In certain respects, he or she functions more like a manager than an engineer. This can come as a rude surprise.

As a young professional engineer, up to this point in your career you have depended mainly on your own skills and abilities to succeed. In high school and college it was you who figured out what the teacher or professor wanted and how to get it done. You burned the midnight oil and worked until you finished an assignment. As a brand-new engineer, you designed the piping or wrote the report or analyzed the stresses in the machine part. You did what it took to get it done when your boss wanted it, or when the project schedule demanded it.

The project engineer must learn a new skill: management. Of course, the project engineering position doesn't have all the dimensions and authorities of a management position, but there are many similarities when it comes to directing the work and dealing with people. At the same time there remains a significant technical component to the job. The project engineer uses engineering skills to bring judgment to decisions and to understand the arguments and rationale that other engineers present.

I've seen many talented engineers struggle with the transition from doing technical work to managing the efforts of others. Their first inclination is to do a problematic task themselves because it takes too long to teach someone else. They soon realize that they are drowning in work. They find themselves tending to details when they should be solving serious technical or commercial problems.

A person's entire success as a project engineer depends on being able to throw a switch in her or his mind. A project engineer must become a manager—and act like one.

This chapter explores the management fundamentals that a project engineer needs. It's a crash course in management, and you have already read the first lesson:

Think and act like a manager.

THE WAY IT SHOULD BE

The functions of management (Figure 3.1) define the role of a manager. They fall into two broad categories:

- The task-oriented side, which often gets most of the attention
- The people-oriented side, which can make the difference between ordinary and extraordinary performance

THE TASK SIDE

Plan

The plan is the foundation of the job. For small jobs, it can be relatively simple. For large jobs it's more elaborate, but it should still stick to the essentials. The plan for the project engineer's area must be in sync with the overall project plan and aligned with the project objectives. As we discussed in Chapter 1, the project engineer's area objectives complement (and fill in the details of) the overall project objectives.

The planning process involves the project engineer's team and other stakeholders. If their input is accepted and respected, they will become committed to the plan. To this end, hold some brainstorming sessions with small groups or in a one-on-one setting.

The steps of the project engineer's planning process are the following:

1. Set objectives. Make sure you have defined specifically what is to be done and when it is needed.
2. Establish that the work is feasible and adequately defined.

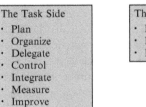

Figure 3.1   Functions of management for project engineers.

3. Check the input data and requirements.
4. Break the work down into activities and assign responsibilities to team members.
5. Assign a duration to each activity and establish the dependency between activities (for example, which activities must be finished before others are started).
6. Estimate the work hours and cost to complete each activity.
7. Develop a baseline cost and schedule estimate that is aligned with the project's cost and schedule estimate.

Step 1 establishes, in broad terms, what is to be accomplished and its timing. It guides your and your team's work. Steps 2 and 3 ensure you have a sound foundation for the plan, so that someone won't discover a fundamental flaw in the design basis, later on. The rest of the steps above are planning, scheduling, and budgeting.

Organize

Organizing is an art based on experience. In the broadest sense, organizing involves mobilizing people, equipment, and materials to accomplish a job. In the context of project engineering, it emphasizes organizing the people.

With the activities, risks, and vulnerabilities of the job in mind, the project engineer visualizes the essential workforce needed for control. Technically demanding work and new technology require the surveillance of experts. Interfaces always need attention, so if there are a lot of them, plan accordingly. Contractual problems and design changes call for the occasional help of commercial people. Activities subject to cost or schedule overruns must be watched. The most significant cost and schedule risks require personnel to monitor and control them. People must be put where needed to control the work, so the project engineer looks at the activities and estimates the number and types of people to be on the area team.

Project engineers often draw up a matrix of "tasks versus responsible people." Let's consider an example. Terry is the client's Lead Structural Project Engineer for a large, high-tech job. He has three people on his team and some part-time specialists. To divide the work of his team, Terry prepares a simple spreadsheet with the people listed across the top and the main or recurring tasks down the left side. An "R" in the cells of the spreadsheet designates responsibility for that task. An "I" means that the person is involved, and "C" means that the person must be consulted at some designated point in the task. Sometimes "A" is used to indicate overall accountability on more formal responsibility matrices involving additional layers of management. Table 3.1 shows part of Terry's responsibility matrix.

Larger jobs have an organization plan that consists of organizational responsibilities, organization charts, and job descriptions for each phase of the project. The organization charts portray the key positions and lines of reporting. A simple

Table 3.1

Responsibility Matrix

Personnel	Terry	Shannon	John	Chris	Lynn
			Structural Responsibilities		
Position	Proj Engr	Str Engr	Str Engr	Foundation Engr	Welding Engr
Percentage of Time	100%	100%	100%	50%	50%
Tasks					
Assess contractor's drawing control systems	R	I			I
Check design input	R	I	I	I	I
Validate software		R		C	
Verify main steel		R			C
Verify critical secondary steel			R		C
Verify design of main welds		I			R
Verify foundation design		C		R	
Follow-up clashes holds		I	R		
Review const. drawings	R	I	I	I	I

R = Responsible
I = Involved on a continuing basis
C = Must be consulted at designated points in the task by the responsible engineer

organization chart for the Engineering Phase of a fictitious project is shown in Figure 3.2.

Organizing is essentially establishing who is on the team, to whom they report, and what each person's main responsibilities are. Whether the project engineer uses a responsibility matrix or an organization chart (or both) to organize the team, a kickoff meeting is essential to explain the responsibilities and coordinate the interactions between team members.

Delegate

Here is a directive that may go against the grain of your upbringing or temperament, but it is crucial for your success and the success of the project. *You must ruthlessly and shamelessly delegate the work to others.* At first it may seem that by delegating, you're fighting with your nature as a conscientious engineer, but you'll soon learn how to control things without doing all the work.

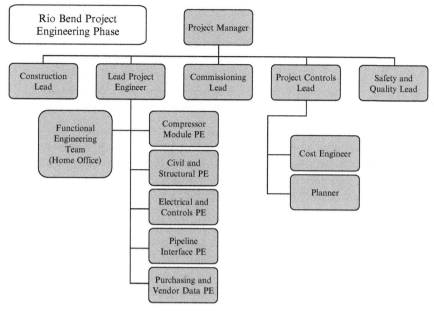

Figure 3.2 Organization chart.

The responsibility matrix (see Table 3.1) is the first step in delegation. It automatically delegates the continuing responsibilities. However, during the job, thousands of items will come across a project engineer's desk that must be handled. These should be assigned to a person on the team who can handle that particular task. The project engineer will want to do some of those items personally because of their importance or the privacy that's required. Let's consider a few examples:

- Your boss informs you that the President of the client's corporation will visit the construction site. He asks you to prepare an agenda for the visit and give a 15-minute overview and safety orientation. Obviously this is an important matter that you choose to handle yourself.
- The Project Controls Manager calls and wants a cost and schedule forecast for your area in the morning. Even though you know most of the information, you delegate the task to the engineer to whom you have assigned the cost and schedule responsibility.
- The Human Relations supervisor informs you that one of your team has been implicated in a minor sexual harassment incident. You're asked to warn the individual of the seriousness of this kind of unwanted behavior. Of course, you can't delegate such a private matter.

- The Project Change Management Board is meeting in four days to review a change to your area's structural design. It may have a large commercial impact. You previously worked on this and had actually run the original design calculations. However, you have assigned responsibility for structural issues to a new member of your team. You decide to delegate the redesign responsibility, but stay in close contact with the work to train the new employee and keep an eye on the commercial implications.

Delegation requires that information be passed to the person who will do the task. Your position as a project engineer will make information available to you that your team doesn't have. If you have regular (weekly) meetings with your team and pass on relevant information, they will begin to have the same understanding and priorities that you have. This will help them to carry out delegated tasks more effectively.

You'll get more work done if you give delegation a high priority. When you arrive at the office each morning, look at your task list and check the ones that can be delegated. You can then delegate those tasks before you start your own high priorities. That way the team is working on the delegated tasks while you're working on yours. You can check with them later to see how things are going or if there are questions.

Implicit in delegating is the need to balance the workload across the team. Suppose that one person on your team is outstanding in many ways, like a football player who can not only pass, but can catch, placekick, tackle, and run the ball better than anyone else. Clearly this person is destined for promotion, but let's suppose she is young and just another member of the team. You certainly have to give her challenging work to keep her motivated, but she can't do everything. If you don't share the workload, your star will be swamped, and the others will resent the fact that they aren't fully participating. Teamwork is a key ingredient in any team, and each member must pull his or her load. You'll have to delegate tasks to everyone, even though it may take some team members longer to do the tasks and require more supervision or training on your part.

Training team members to perform delegated tasks is worth the time it takes. Team members can't do what you would do the first few times, so take time to explain what's needed and when it's needed. You'll generally be pleased with how quickly most people improve. Trust is the grease that improves their performance and builds confidence. You may also be surprised to learn that people are motivated by being asked to help.

That having been said, it's always wise for the project engineer to check the results from a delegated task before sending them up the line. The extent of the checking will depend on the importance or criticality of the information and certainly to whom it's being sent. For example, you wouldn't want to give your boss a cost report that one of your team members had prepared if it contained wrong information. Carry out some simple due diligence to determine if the information is complete and correct.

And—to continue our football metaphor—someone may fumble the ball each time you hand it to him. It's always possible that some team members can't or won't perform the tasks they are given. When a pattern of poor performance develops, take it up with your supervisor. As a project engineer, you most likely won't have authority for personnel matters, but your supervisor will. This is called *upward delegation*. You have to be careful about delegating tasks upwardly, since delegation is supposed to flow the other way. Your supervisor wants you to handle your responsibilities. In cases like this, however, when you don't have the necessary authority, it's proper to delegate the matter to your boss.

Control

For the most part, control involves finding what's wrong and fixing it. Control is the process of making sure that all the project's requirements (safety, quality, cost, schedule, interfaces, and more) are being met. To control, the project engineer must have a baseline plan or target to measure against. He or she will then deal with deviations from that plan—safety incidents, cost overruns, schedule delays, quality nonconformances, major design changes, interface mismatches, and the list goes on.

The case study in Chapter 5 illustrates practical examples of controlling project work. A case study is probably more illustrative than going into detail on sophisticated control techniques. Project engineers who rely too heavily on control techniques and concentrate too little on the substance of what's being done will sometimes miss serious problems. You have to look at the work behind the cost and schedule numbers to understand the reasons why costs are so high or the work is taking too long.

Management of changes and management of interfaces are important control mechanisms, in addition to the normal safety, cost, and schedule control processes. Surveillance, which includes quality control, is also an important part of controlling the work in your area. Your project team will have systems for all of those. Just learn what they are and keep in step.

Integrate

Integration is the function of linking the work in a project engineer's area with the rest of the project. The project engineer is the linking pin between the people who look to him or her for leadership, and the project management, other project engineers, functional departments, contractors, and vendors. The project engineer is also the most important source of information for the people on the team.

Interface control is a good example of the integration function at work. As you remember from Chapter 2, the project engineer has the primary responsibility for the interfaces with her or his area.

Integrating doesn't mean that the project engineer becomes a bottleneck or turns the team into an information silo—working in their designated area in isolation from what's happening on the project. On the contrary, the team must communicate and be connected to the rest of the project. The project engineer is the facilitator for the team's communication with other organizations and the team's spokesperson on important matters. When a team member gives out incorrect or inappropriate information, the project engineer steps in (in a positive, nonjudgmental way) to rectify the situation or control the damage.

Measure

Measurement is the acquisition and consolidation of statistics that help the project and others understand how the job is going. Progress on safety, quality, cost, and schedule must be measured before it can be controlled. It's difficult to know what to control, without measuring the information that will bring problems to the forefront. For example, the project engineer and his or her team will keep track of the cost of each activity in the plan and compare the current costs with what is budgeted. They and others report some of this information to the project control team so that the Project Manager can gauge if the cost objectives are being met. Sometimes increases in one area balance decreases in others, and no action is necessary. However, if significant cost overruns appear or other adverse trends come to light, the project management is informed and can take action to bring costs back in line with the budget.

The project team also measures information that helps them make judgments about what control measures are needed to fix problems. The quality measurements mentioned in Chapter 2 are examples. When excessive defects appear in a product, team management will investigate root causes of the trouble and improve the work processes.

In the field of safety, the project will monitor leading indicators that give warning of serious lost-time or fatal accidents. By measuring near misses, first-aid cases, and other minor safety incidents and taking corrective actions, more serious accidents can be avoided.

In addition to the normal measuring and reporting, early, frank, and candid communication of events and problems to your boss is a sound strategy—but don't overdo it. Most bosses, unless they are micromanagers, will want their subordinates to handle their own responsibilities. Thus, the project engineer only needs to communicate information that is important to the overall project or is needed by the boss to do his or her job. If you bring up trivia, you risk a couple of undesirable outcomes. One could be that your boss gets that look on her face,

"Why are you wasting my time with this?" That will reflect on your judgment. Another response could be that your boss becomes engaged in this trivial matter, which he knows nothing about. He could give you off-base advice that you are now obliged to implement. Bringing up trivial or irrelevant information to your boss is high risk and seldom in your best interest.

Improve

Every effective management system has a procedure for improving itself. Organizations learn from what happens and improve their plans, work processes, business controls, or even the organization. Sometimes when faced with setbacks, the project must develop a recovery plan that will change the cost and schedule baselines.

Many project engineers are so busy with their work that they don't take the time or effort to improve the way they do things. The end of a major phase or the end of the project is a good time to get the team together and review the lessons that they have collectively learned. In many organizations, this procedure is required by company policy.

Another sure way of improving an organization is to manage its human resources and tend to the needs of its people. Let's look at that next.

THE PEOPLE SIDE

The most effective managers are able to balance accomplishing their tasks with taking care of their people. They keep people informed. They get people engaged in the work and onto the playing field. They create an environment that allows people to participate in decision making and be empowered to take action. They lead them through difficult times with confidence and take pressure off during stressful situations.

The ability to influence people lies at the heart of management and leadership. These two skills are intertwined; thus it's useful to define them (Hersey and Blanchard, 1982):

- Management is "working with and through individuals and groups to accomplish organizational goals."
- "Leadership is the activity of influencing people to strive willingly for group objectives."

Working with and through others requires interpersonal skills if a leader wants to have influence:

- It's about the way you communicate with people.
- It's about the way you respect their work contributions.

- It's about listening and how you react to their comments and suggestions.
- It's about trying to understand what motivates their behavior in a given situation.

For some of you, interpersonal skills will come easily, and you can rely on your instincts. For others it may take some work. But it's always useful, as a part of your career development, to seek training in interpersonal skills and basic management skills. This is because the "people side" is so important. The way you manage people makes the difference between poor performance of your group and ordinary performance. It's also the difference between ordinary and extraordinary performance.

Now let's examine three aspects of management that project engineers will find useful as they try to influence people.

Motivate

Motivation inspires individuals to work effectively. It is complex, but three qualities tend to strongly influence it (Hersey and Blanchard, 1982).

Self-Esteem or Power

People are motivated when they feel *self-esteem* and confidence from what they do. Also in this category is the feeling of being in control or having power. If a project engineer interacts with employees in a positive way that reinforces their self-esteem and confidence, it tends to motivate them. (A negative approach will tend to demotivate.) Putting people in positions of responsibility or trust is also a motivator that is based on self-esteem. As mentioned earlier, delegation of tasks in a positive way can be motivational.

Self-Actualization or Achievement

Many people are motivated by feelings of achievement about things they have accomplished themselves, which is called *self-actualization*. This type of motivation comes from within the individual, if she or he believes they have accomplished something important and significant. One of a manager's roles is to assign employees to jobs where they can work to their potential and achieve major accomplishments. Self-actualization is reinforced by recognition, which relates to self-esteem. The recognition can be as simple as a complimentary remark or something more formal like an announcement or recognition ceremony, but in either case it must be genuine.

Affiliation

People can also be motivated by feeling that they are part of a group. Much of this feeling comes from enjoying the association with fellow workers. Part of it is pride in what the organization is accomplishing, and the satisfaction of feeling that working toward group goals is fulfilling the employee's own goals. For some, particularly people involved in projects, *affiliation* with their group is a strong motivator that can be reinforced through team building.

Build a Team

It is generally accepted that a team's results are better than the sum of the individual results. When employees work as a team and exhibit good team process, synergy occurs. The project engineer can promote this type of performance through team building.

Team building should be a planned, deliberate process that starts at the team's kickoff meeting. At that meeting, the project engineer can establish a positive tone and set the expectations for how the group should interact among themselves and others. As the team works on early tasks, the leader can facilitate the kind of behavior that promotes effective interpersonal relationships and good group process. In other words, the project engineer should concentrate not only on *what* the group is doing, but also on *how* they are performing as a group.

Every group, including your team, goes through four stages—forming, storming, norming, and performing. In the forming stage, the group members are getting to know each other, and it's crucial for the leader to maintain a supportive environment where everyone's ideas are respected. As the group begins to work on their tasks, issues between group members or with the leader will begin to emerge. As discussion continues, disagreements may occur over some of the issues and, in some cases, conflicts can arise. This is where storming comes in. With proper facilitation, group members share their viewpoints and begin to move into the norming phase. Members develop trust as they discuss the issues and form common opinions. As this process develops, the group moves to the performing phase where they work as a productive team. If the climate within the group isn't supportive, it's possible that the group may never get out of the storming phase. If the group has strong personalities involved, and the mission is crucial (like reengineering a company's business processes) it may be prudent to hire a professional facilitator to help the group work through the phases and become productive.

Another part of team building is creating direction—goal setting. You will be amazed at what people can accomplish if they have a vision of what's required and if they understand the importance of what they are doing. And when they share in creating that vision, their motivation is even stronger.

It's advisable to hold team-building sessions prior to a new phase of the project. New people may be joining the team and new tasks will be introduced. For example, before starting the purchasing phase, it's good practice to get the engineers, buyers, engineering and procurement supervisors, and other relevant people together to discuss, plan, and coordinate the work. Team building with the key people who must work together on interfaces can also be valuable time spent. Social activities (a lunch or dinner after the meeting) as a group can promote team-building objectives.

In a nutshell, team building will build a team identity, develop team spirit, improve the possibility of outstanding team performance, and enhance each individual's *commitment* to the project.

Develop People

Project engineers won't generally have a supervisory responsibility for the career management and personal development of the people who work for them. But that doesn't mean they don't play a part in developing those employees, who look to them for leadership.

When assigning tasks, the project engineer should try to place people in jobs for which they are qualified. If necessary, give them on-the-job or formal training. To the extent possible, team members should be given tasks that will challenge them. As they do the work, the project engineer should coach and counsel them in a positive way. Giving timely advice and recognizing their improvements will foster the employees' development and motivation. As the project winds down, the project engineer can talk with the appropriate supervisors about moving employees to positions of more responsibility if they are ready for it.

THE WAY IT IS

So, that's the rational or cerebral face of management, but is that the whole picture? Henry Mintzberg (1990), in his award-winning article, "The Manager's Job: Folklore and Fact," reveals what it's really like to manage.

Mintzberg developed a more practical and useful view by digging deeply into extensive, credible research on all types of managers—"foremen, factory supervisors, staff managers, field sales managers, hospital administrators, presidents of companies and nations, and even street gang leaders" (Mintzberg, 1990). They all experienced the same thing. They "work at an unrelenting pace; their activities are characterized by brevity, variety and discontinuity, and they are strongly oriented to action and dislike reflective activities" (Mintzberg, 1990).

This is a somewhat uncomplimentary view of management. Yet, it's amazing that over the years managers have commented to Mintzberg (1990): "You make me feel so good. I thought all those other managers were planning, organizing, coordinating and controlling, while I was busy being interrupted, jumping from one issue to another, and trying to keep the lid on chaos."

One of the lasting values of Mintzberg's work (1990) is his recognition that there are two sides to management: "there has to be a balance between the cerebral and the insightful." The cerebral side is planning, organizing, delegating, controlling, integrating, measuring, and improving. Insightful management is more elusive. It involves absorbing and assimilating direct as well as peripheral information (mostly through conversations), and it "depends on direct experience and personal knowledge that comes from intimate contact" (Mintzberg, 1990). It's this insightful side that drives good managers and makes their job so hectic—and so interesting.

Mintzberg (1990) put the two faces of management together in his set of manager's roles (Figure 3.3). By virtue of being in charge of an organization or unit, the manager has certain formal authority and status. From this Mintzberg claims, come three *interpersonal roles*:

- Figurehead—acting as a representative or symbol of the unit
- Leader—influencing and directing the work of the unit
- Liaison—interacting with peers, subordinates, higher management, and important people from other organizations

The interpersonal relations from these roles give the manager access to a wealth of information and enable the manager to carry out *informational roles:*

- Monitor—collecting and processing information from the manager's own unit, as well as outside organizations
- Disseminator—passing on some privileged information to subordinates

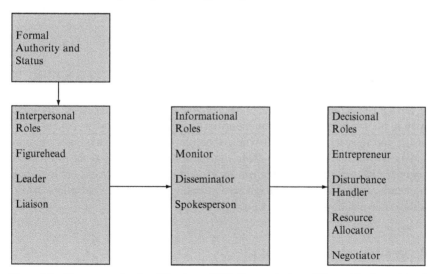

Figure 3.3 The manager's roles (Mintzberg, 1990). By permission of Harvard Business School Press. From "The Manager's Job: Folklore and Fact" by Henry Mintzberg. Boston, MA 1990, pp. 1–36. Copyright © 1990 by the Harvard Business School Publishing Corporation; all rights reserved.

- Spokesperson—sending some information to people outside the manager's unit to peers, management, and outsiders

Information from the interpersonal roles provides the input for the manager's *decisional roles*, which unfold when the manager has the authority and the most complete set of relevant information needed to make organizational decisions and set strategies:

- Entrepreneur—adapting to changing environments, identifying opportunities, developing new products or initiatives, improving the unit
- Disturbance handler—solving serious incidents, conflicts, or problems that arise
- Resource Allocator—appropriating funds to projects and endeavors, assigning key personnel, allocating equipment and materials, determining the effect of one decision on the other decisions
- Negotiator—persuading and influencing others (both internal and external) to align with the manager's organizational objectives

Are all of these roles relevant to the project engineer? No. Mintzberg's work covers every level of management from CEO to shop supervisor. Also, his approach tends to deemphasize planning. Project work requires that leaders take time to plan. It's part of the job that happens at the onset. Projects fail if they aren't planned well. You'll have to sift through Mintzberg's roles and find the nuggets of truth that fit your own situation. The exercise is worth your while.

MANAGEMENT SKILLS FOR A PROJECT ENGINEER

Let's take a moment to consider how a project engineer should apply this information. It's useful to keep the functions of management in mind as a checklist to ensure you're accounting for the whole spectrum of management considerations. *To ensure you've covered everything as you approach a task, think over: plan, organize, delegate, control, integrate, measure, and improve.* Occasionally take a moment to reflect and ask yourself some questions regarding your relationships with people and how the work is going.

PEOPLE-RELATED QUESTIONS

Since your formal authority as a project engineer is limited, you'll have to rely on your interpersonal skills to influence your team and others. (There's more on this in Chapter 6.) Ask yourself:

- Am I balancing my emphasis between the task and the people, especially during difficult or stressful situations?
- Do I motivate the team members by recognizing their accomplishments, giving credit where credit is due, assigning challenging work, and promoting a team spirit?
- Do we work as a team and take time to do team building?
- Do we evaluate our performance as a team?
- Do I take steps to develop the people on my team?
- Am I keeping my boss informed?

TASK-RELATED QUESTIONS

While you're leading the team in their day-to-day tasks, remember Mintzberg's perspective that management is challenging, hectic, and chaotic. But don't let that deter you from performing the leadership functions that a project engineer must do. Set priorities so that your team concentrates on the important tasks, and again ask yourself some questions:

Plan

- Do we have the right plan and are we sticking to it?
- Are we working on our top priorities?

Organize

- Does my team have the necessary full- and part-time skills?
- Have I clearly established and communicated the responsibilities to the team?
- Do they understand their responsibilities?

Delegate

- Am I taking the time to delegate?
- Am I prioritizing the tasks that need to be delegated and handling those first?
- Am I allocating work to the whole team?

Control

- Are we controlling safety, quality, cost, and schedule?
- Do we use sound, ethical business practices?

Integrate

- Am I linking the team with management, my peers, and the rest of the project?
- Do I facilitate communication and information flow between our area and others?

Measure

- Are we recording the necessary information to facilitate control and fulfill our reporting requirements?
- Do I have enough information to make decisions?

Improve

- Do we occasionally take time to learn from our successes and mistakes?
- Are we putting those lessons into practice?

Remember that you're the leader and have total area responsibility. That gives you a mandate to coordinate all that goes on within the boundaries of your area. Look to your role models for guidance and example, then think and act like a manager...even if you're not one quite yet.

REFERENCES

Hersey, P., and Blanchard, K. H., *Management of Organizational Behavior: Utilizing Human Resources* (4th ed.) (Prentice Hall, Inc., New Jersey, 1982), pp. 3, 14–82.
Mintzberg, H., "The Manager's Job: Folklore and Fact," *Harvard Business Review on Leadership* (The Harvard Business School Press, Boston, 1990), pp. 4, 12–21, 29–31.

Chapter 4

How Projects Work

PLAN THE WORK AND WORK THE PLAN

Robert M. Clawson, one of the great project managers of the North Sea oil and gas construction boom, often used an expression that captures the headlines of project engineering. Bob would tell his engineers, "Plan the work and work the plan." Perhaps the saying even pre-dates Bob, but it's still sound advice today.

If we relate that saying to the functions of management in Chapter 3:

- "Plan the work" involves planning, organizing, and most of delegating.
- "Work the plan" is the day-to-day part of delegating, plus controlling, integrating, measuring, and improving. It's also the people side—motivating, building a team, and developing people.

Bob's expression implies having a sound, detailed, well-thought-out plan, and then following it. After "work the plan," Bob added, "No zigs...no zags." This means to focus on working the plan without any unnecessary detours or side-tracks. Of course, things will happen and changes will be necessary to adapt to new situations, but, in general, follow the plan with commitment. Doing otherwise on large projects can lead to confusion and trouble.

OVERALL PROJECT FRAMEWORK

If we view a large project from an *experienced project manager's perspective*, it's a set of related plans and activities to accomplish certain specific objectives.

It's "plan the work" and "work the plan" on a broad scale. Successful project managers know that large projects require structured activities, executed in a particular order.

When viewed from an *inexperienced project engineer's point of view*, a project is a torrent of activity rushing along, sweeping away everything in its path. It's natural for most new engineers to find security in the sheltered waters of their own job, put their head down, and focus on the here and now. They ignore what's happening in other areas, or what's coming down the river toward their own area. However, with an understanding of the overall project framework, project engineers are able to anticipate the tasks ahead and plan for them. Understanding how projects work helps them put their jobs into broader perspective and avoid being blind-sided.

MAJOR PHASES

Figure 4.1 illustrates the major phases of a large project and how they play out over the project's life. Near the center is a downward-pointing arrow labeled "Project Approval." Activities to the left of that milestone are dedicated to *planning*, in preparation of the decision to start or fund the project. The goal is to optimize the project plan—to make it the best project it can be under the circumstances and constraints that may exist.

The major phases within planning are *evaluation* of the project (measured against some prescribed criteria) and *definition* of the project's basis (design criteria, technical concept, execution plans, schedule, and cost estimate).

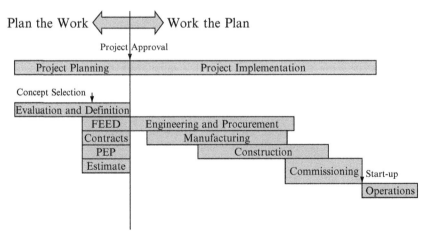

Figure 4.1   Overall project framework.

Activities to the right of the "Project Approval" arrow are *project implementation* (executing the plan, or actually doing the project). Project implementation has four distinct phases:

- Engineering and procurement
- Manufacturing
- Construction
- Commissioning and start-up

On the far right of Figure 4.1 is a bar called "Operations." This element is not strictly part of the project, but, during the early part of operations, a remnant of the project team often works closely with the operational personnel running the new facility to make sure it works as it should.

Successful project teams develop a sound, well-defined basis during planning. When they cross the great divide into implementation, they stop optimizing and tinkering with the plan or the concept. Instead they concentrate on getting the job done and solving any problems that get in their way.

PROJECT MANAGEMENT

A crucial factor for project success is having someone in charge—a single project manager, responsible for the entire project and accountable to the management of the venture for safety, quality, cost, progress, and all the other project objectives. The project manager establishes clear lines of responsibility within her or his project management team for all aspects of the work. On large projects the project manager will create overlapping responsibilities, such as

- engineering versus commercial,
- engineering versus construction, and
- overall surveillance of engineering, procurement, and construction by the quality manager, project controls manager, or safety, health, and environmental manager.

Those checks and balances give the project manager independent, unfiltered information when trouble is brewing. They essentially make the project self-correcting.

Some companies employ a matrix approach, where the project manager draws resources from a functional organization. For example, the engineering is performed under the supervision of the company's engineering manager and not as part of a dedicated project team that reports to the project manager. The matrix approach can be cost-effective and may work for small projects if there are processes in place to clearly define the responsibilities. But the matrix approach often breaks down in a barrage of finger pointing when things go wrong.

PLANNING: EVALUATION AND DEFINITION

EVALUATION

The main purpose of evaluation is to make sure that the project is viable. But what does "viable" mean? As with many questions, the answer depends on who you ask. An executive in a private company might say viability means that the return on the money invested in the project meets the firm's investment criteria and is attractive compared to other opportunities in which they might invest. Low risk, technical feasibility, environmental friendliness, and on-time completion could also be important objectives. The project manager for a large public sector project (like a major bridge or highway) might also say that viability means that the project is economical, but he or she would have in mind the value to the community, and not profitability. Factors such as opening up new opportunities, political viability, environmental acceptability, and the minimization of disruption or change could be valued more highly than present-value economics.

Evaluation is the creative part of the project during which the business planning is accomplished and the design concept is selected. The client's evaluation team develops a planning basis, coordinates the planning effort, and generally screens a broad range of concepts. The concepts are technically defined to a level that allows feasibility to be established and concept risks (risks inherent in a given concept) to be assessed. Unfeasible or unsafe concepts are eliminated from further consideration.

Screening-level costs and schedules are estimated for the feasible concepts. Those concepts are then evaluated and screened (based on economic parameters and all other relevant considerations, like those mentioned previously). The best concepts are short-listed for further evaluation, while the others are eliminated.

The short-listed concepts are technically defined in more detail and further optimized by reducing cost, improving schedule, enhancing output of the facility, and a host of other improvements. At this idea stage of the project (before the detailed engineering, procurement, and construction have started), improvements yield the greatest benefits. It's nearly always worth taking the time to optimize a project during evaluation. Also, as competing concepts become more defined, the team identifies, evaluates, and finds ways to mitigate the risks associated with each remaining concept.

After several cycles of evaluation, the client's management *selects the final concept.* In each cycle, fewer concepts are considered, and the technical definition of each concept increases. The selected concept is, therefore, well defined and robust. A robust concept has the inherent flexibility to sustain its viability in the face of any reasonable challenge that may develop during the project's life.

DEFINITION

After the concept is selected, the focus of the client's planning team shifts from evaluation to definition. The planners are still doing evaluation, but at this milestone the engineers step forward and start four distinct sets of definition activities:

- Front-end engineering design (FEED) that defines the chosen concept in enough technical detail to bid the job
- Preparation of the contracting strategy and bid documents needed to engineer, procure, and construct the facility
- Planning how the work will be executed and how long it will take
- Estimating what it will cost

Let's consider each of these in a little more detail as we develop the sequence of events leading up to project approval and contract awards.

Front-End Engineering Design (FEED)

The client's project team manages front-end engineering design. However, a few key contractors are usually brought on board to do the FEED. The client's and FEED contractors' engineering teams work together to prepare the technical drawings and documents that describe the job in enough detail to bid it. The client specifies how the work during implementation will be arranged into contracts, so that the FEED engineering contractor can focus on preparing the right technical documents (FEED deliverables) to go into the contracts. Some clients try to do the FEED engineering in-house, but it's usually best to employ the detailed engineering expertise of a FEED contractor to develop the details properly. The major FEED activities include the following:

- Develop functional requirements for the facility and its equipment
- Complete the design basis (design temperatures, wind speed, soil conditions, or whatever other considerations apply)
- Prepare a detailed scope of work for the contractors to follow
- Define the configuration of the facility and the major systems
- Review and complete the client's specifications and develop a design specification for detailed engineering
- Conduct safety studies and assess risks
- Define major interfaces
- Efficiently produce the drawings and documents

If the FEED contractor has engineered and constructed major projects, the FEED deliverables will reflect that experience and should lead to high-quality bid documents—it takes one to know one!

Contracting Strategy

The nature of bidding in all project phases depends on the client's contracting strategy established during definition. It's generally approved by high levels of management. Examples of a few popular contracting strategies are explained below. A handy way to quickly convey a contracting strategy is illustrated in Figures 4.2, 4.3, and 4.4. Listed across the top of each of those contracting strategies are the main components or tasks of a hypothetical project. Those are the work to be done. Often the project's organizational structure is aligned with the main components and tasks. On large projects, such as the one shown here, major related components and tasks are grouped into subprojects headed by a subproject manager. Subprojects function as projects within a project to provide better accountability when hundreds of millions of dollars are being spent. Down the left side of the contracting strategy are the project's phases, running from the client's management at the top to commissioning and start-up at the bottom. Those are the project phases from beginning to end, respectively.

EPC

Some situations favor an engineering, procurement, and construction (EPC; see Figure 4.2) contracting strategy. Here the contractor executes the lion's share of the work with varying degrees of client intervention. Some clients throw the entire job over the fence to the contractor, while others maintain substantial control through the presence of large site teams. Clients favor an EPC strategy when the project is similar to other recent projects, and there are several experienced contractors to choose from.

	Project Components or Subprojects			
Phases	Civil & Foundation	Plant	High-Tech Equip	Offsite Utilities
Client Project Management	Client Project Management Team (PMT)			
FEED		FEED Contractor		
Project Management	EPC Contractor			Client PMT
Engineering			High-Tech Proprietary Equipment Supplier	
Procurement				
Construction or Manufacturing				
Commissioning				
Start-up	Client Start-up Team			

Figure 4.2   Engineering, procurement, and construction (EPC) contracting strategy.

E&P Contractor

Other situations may indicate the need for an engineering and procurement contractor (E&P contractor; see Figure 4.3) to maintain close client control over the engineering and purchasing. The client then bids the construction contracts with the help of the E&P contractor. This approach is used for high-tech or other projects requiring considerable engineering definition and control. The client can influence the design, manage the purchasing and contracting, and control the major engineering, procurement, and construction interfaces. When problems occur at those interfaces, the client receives direct information from at least two rival contractors, rather than observing the problem from outside, as would be the case with an EPC contract.

Phases	Project Components or Subprojects			
	Civil & Foundation	Plant	High-Tech Equip	Offsite Utilities
Client Project Management	Client Project Management Team (PMT)			
FEED		FEED Contractor		
Engineering	Engineering and Procurement (E&P) Contractor		High-Tech Proprietary Equipment Supplier	Client PMT
Procurement				
Construction or Manufacturing	Construction Contract A	Construction Contract B		
Commissioning	Client Commissioning Team			
Start-up	Client Start-up Team			

Figure 4.3   E&P contractor contracting strategy.

EPCM

A third contracting strategy, engineering, procurement, and construction management (EPCM; see Figure 4.4), strengthens the role of the contractor in bidding, supervising, and administering the construction. The EPCM contractor assists or leverages the client's project team in bidding the construction contracts. However, the client retains control over the selection of the construction contractors. Clients like an EPCM approach when they want high involvement in the project but don't have enough experienced people to manage the job themselves.

There are many other contracting strategies and variations, but this will give you an understanding of the main ones. They drive how the work is divided into contracts.

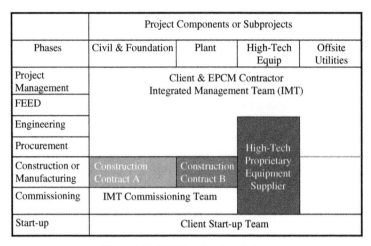

	Project Components or Subprojects			
Phases	Civil & Foundation	Plant	High-Tech Equip	Offsite Utilities
Project Management	Client & EPCM Contractor Integrated Management Team (IMT)			
FEED				
Engineering				
Procurement			High-Tech Proprietary Equipment Supplier	
Construction or Manufacturing	Construction Contract A	Construction Contract B		
Commissioning	IMT Commissioning Team			
Start-up	Client Start-up Team			

Figure 4.4   EPCM contracting strategy.

Bid Documents and Contract Documents

As you might expect, the bid documents are almost exclusively assembled by the client's team to maintain confidentiality and security in the bidding process. The bid package generally consists of an Invitation to Bid (that requests from the contractors considerable commercial, technical, and project execution information) along with a draft of the Contract Terms and Conditions. Contractors can't assess the contractual risks without a copy of the proposed contract. A number of exhibits such as the scope of work, the format for paying the contractor, project milestone dates, administrative requirements, and coordination and control procedures are attached to the Invitation to Bid. The client's design basis and selected FEED deliverables (specs, drawings, studies, and other data) are also attached to technically specify the job. The client's Project Execution Plan and schedule are often included. Other exhibits set the financial, quality, safety, environmental, and other standards that the contractors must live up to. Well-planned projects have two characteristics:

- They develop a matrix that defines the important interfaces between contracts.
- They establish procedures to control those interfaces.

Project Execution Plan

In parallel with FEED and the contracting effort, the client's project team prepares a plan to implement the project. Often called the Project Execution Plan

(PEP), it covers all aspects of the job. The project team develops organization plans, schedules, and control measures for engineering and construction. They prepare quality plans, interface plans, and safety plans. They develop contracting, purchasing, and logistical plans. Where necessary, all these plans and control measures find their way into the major contract documents, so that contractors can implement them. After all, it's the contractors and vendors who do nearly all the tangible work on projects. The project's baseline schedule grows out of this planning effort.

Cost Estimate

The cost estimate prepared at this point is of crucial importance, whether for private or public sector projects. It serves as the budget for funding the project and the baseline for controlling project expenditures. Most good projects build a detailed estimate, based on engineering definition and the PEP. They benchmark or compare their estimate with relevant data from similar projects to verify the estimate's accuracy.

Contractors' Bidding

When the contractors receive the Invitation to Bid, they review it and prepare a detailed outline of their bid proposal with assigned responsibilities for preparing each section. Part of the proposal team reviews the client's basis and seeks clarification where necessary. Normally the contractor must supplement the client's basis when preparing the bid.

The team then prepares a detailed work breakdown structure (WBS) that includes every significant part of the contractual work. The WBS is the basis for planning the job. It is utilized to write a comprehensive scope of work and estimate the weight and cost for each item in the WBS. Work-hours and durations are estimated for each WBS item, and a networked schedule is prepared for engineering, procurement, construction, commissioning, and start-up. Most of the proposal team participates in preparing a technical description of the work and a contractor's PEP that will be submitted with their proposal.

Bids or quotations and delivery schedules are solicited from suppliers of equipment, bulk materials, and subcontracted work. The more firm those quotes are, the less risk the contractor must price into the bid. As the suppliers' prices become firm, the commercial proposal is assembled and reviewed by the project management and higher. A comprehensive assessment of technical and commercial risks guides the management's judgment when adding risk money on top of normal contingencies and allowances. The profitability of the job hinges on the quality of the commercial proposal.

Bid Evaluation and Award

The client's teams that prepare the bid documents generally form the nuclei of the client's bid evaluation teams. There is usually a bid evaluation team for each major contract to be bid. Those bid evaluation teams analyze the contractors' bid proposals and recommend a winner. Negotiations continue with the winning bidder for each contract until there is a meeting of the minds. As agreement is reached, the various parts of the bid documents evolve into the signed contracts.

A key factor for project success is a robust bidding process that gives contractors a sound basis to bid on and allows them sufficient time to

- understand the bid documents,
- develop a scope of work, technical proposal (which often requires some engineering), and an execution plan,
- secure commercial commitments from subcontractors and vendors, and
- estimate the cost and mitigate risk.

PROJECT APPROVAL

When the project passes the project approval milestone, the client's planning team has done their job if they've achieved the following:

- A robust concept that can be adapted to the changes that will come
- Sound engineering definition
- Well-formulated contracts that are negotiated and ready to be signed
- A robust Project Execution Plan
- High confidence in the cost and schedule estimates

What the team does in planning the project is crucial to its success in implementing it. Following the project's approval, the activity and expenditure levels ramp up exponentially, as contractors and vendors are mobilized. The planning is finished, and the project implementation begins with engineering and procurement.

PROJECT IMPLEMENTATION: ENGINEERING AND PROCUREMENT

My Project Executive once gave a simple illustration that put my job, as Engineering Manager on a $3.4 billion project, into its most basic terms. He asked me to imagine a worker with empty hands. That worker, he said, needs

- drawings and instructions that tell him or her what to make, and how to make it,

- tools and equipment to carry out the tasks efficiently, and
- building materials, parts, equipment, and other basic components to assemble or manufacture into the final product.

It's the project team's responsibility to get all of those into that worker's hands, when needed. Most of that responsibility rests on the shoulders of the engineering and procurement teams. The project engineers coordinate and control those efforts at the workface (where the work is actually done) to ensure that their teams successfully deliver what the worker needs.

ENGINEERING AND PROCUREMENT—AN INTEGRATED PROCESS

I wish we could separate engineering and procurement into two separate topics, but we can't. It just doesn't work that way. Those are such integrated processes that they must be carried out in concert by the project's engineering and procurement teams. The engineers must tell the procurement people what to contract for or what to buy. Engineers also provide all the technical details that accompany the contracts and purchase orders. The procurement people own the contracting and purchasing documents, and the commercial processes that are used to prepare them and bid them to subcontractors and vendors. That complementary technical and commercial tension between the engineering and procurement teams helps them jointly achieve their shared objectives:

- Design a facility with acceptable safety, environmental and health standards, quality, cost, and schedule
- Procure all the goods and services needed to design, construct, and commission the facility

Design—the first of those objectives—is self-explanatory to engineers, but procurement requires some explanation. It consists of two major activities, purchasing and contracting.

Buyers on the purchasing team write purchase orders to procure manufactured equipment (e.g., compressors, pumps, motors, switch gear, control systems, air conditioners) and bulk materials (e.g., steel, concrete, pipes, wires, valves). Purchase orders can be for items as large as a power generator and as small as bolts, valves, or pipe fittings. Logistics are often included in the purchasing responsibilities except in areas, such as Siberia or Central Africa, where logistics would require extraordinary efforts. In those locations, a logistics contractor, familiar with working in the region, may be hired to manage the movement and storage of the project's equipment and materials.

Contracting specialists work on teams that formulate, bid, evaluate, award, and administer contracts. Contracts are generally for large components that require substantial engineering, procurement, and construction. For example, the civil work to prepare a plant site for construction could be contracted.

Table 4.1 summarizes the main detailed engineering and procurement activities and how they relate. The shaded activities require cooperative interaction between engineering and procurement. Refer to Table 4.1 as you read through the following sections.

Design Basis Verification, Engineering Planning, and Implementation of Quality Management Systems

Detailed engineering starts with verification of the client's design basis. In parallel, project engineers work with design engineers to prepare all the other basis documents that engineering needs. It's usually necessary for contractors and vendors to clarify and supplement the client's basis documents to create the starting point for detailed design.

Another vital part of project initiation is preparing a detailed scope of work—a narrative that summarizes all the tasks to be completed under the contract. This document was started during the bid preparation but now must be updated to reflect

Table 4.1

Engineering and Procurement: An Integrated Process

Engineering Activities	Procurement Activities
Design basis verification, engineering planning, and implementation of quality management systems	Purchasing and contracting strategies and plans
Systems engineering	Purchase orders formulation, technical requisitions, and PO bidding
Manufacturing and quality surveillance	Manufacturing and quality surveillance
	Vendor data receipt
Area engineering—drawings, data, and documents for construction bids	Construction contracts formulation, bidding, evaluation, and award
Area engineering—construction drawings, data, and documents; verification of critical deliverables	Equipment and materials deliveries to construction sites
Commissioning and operating procedures	
Follow-up engineering and as-built documentation	
	Contract administration and changes to contracts
	Contracts and PO closeouts

negotiations and expanded. Effective project management teams prepare a 60-day or 90-day schedule, which lays out the project initiation tasks and deliverables. This gets the team focused on initiating the project and coordinates the activities during that turbulent time.

The engineering contractor's planning begins with a detailed work breakdown structure (WBS). Work breakdown structures can have a variety of formats, but the most effective ones are patterned after both the client's and the contractors' contracting strategies and reflect the end products. This means that, from the beginning, the engineering team focuses on the end products—the technical drawings, data, and documents that

- go into the purchase orders (POs) and contracts,
- go to the construction sites or commissioning teams, or
- are needed to operate the final product.

The engineering and procurement project teams plan all the detailed tasks, according to the WBS. For example, the preparation of individual drawings, specifications, and technical requisitions are reflected in the most detailed project schedules.

During this period, engineering implements the design management systems (needed to ensure accuracy and completeness) and all the other quality management systems (change control, document control, cost and schedule control, and other business controls).

Purchasing and Contracting Strategies and Plans

In parallel with engineering's start-up, the procurement team leads the preparation of the purchasing and contracting strategies and plans. This is the time when the project management decides which purchase orders and contracts will be sent out for bids.

Similar manufactured items are grouped into purchase orders to reduce the amount of purchasing effort—fewer purchase orders results in less work by the purchasing team. The final purchasing plan includes

- the overall purchasing strategies,
- a list of purchase orders to be bid,
- preferred vendors for each purchase order (bid lists),
- standardized timetables for the purchasing activities,
- schedules for the bid-evaluate-award cycles for each purchase order, and
- guidelines and coordinating instructions.

Eventually other details, such as bid-check estimates (the project team's estimate of the price that the lowest bidder will submit), are added as the project progresses.

The fabrication or construction effort is divided into contracts. Effective projects develop detailed scope of work documents in the contracts to explain, as precisely as they can, what the contractor must do. Scopes of work are tailored to obtain the best value from the bids. Large scopes of work reduce the project teams' workload, but also reduce the numbers of bidders—and possibly competition. The project should always tailor the contracts to the industrial base that supports the project. In some countries, political factors influence the preparation of purchasing and contracting strategies and plans. The final contracting plan contains

- the overall contracting strategies and work breakdown structure for the major contracts, including the pricing format (reimbursable, lump sum, schedule of rates),
- a list of all contracts to be bid,
- preferred contractors for each contract (bid lists),
- schedules for the bid-evaluate-award cycles, and
- guidelines and coordinating instructions.

The contracting plan is a living document that evolves over the project.

Procurement personnel interact with their engineering counterparts to ensure consistency in the strategies and schedules. Models or templates for purchase orders and contracts are assembled to guide future purchasing and contracting teams when preparing bid documents.

Systems Engineering

Detailed engineering begins in earnest with systems engineering. Systems (e.g., power, heating and air conditioning) were defined during FEED but are designed during systems engineering. The discipline engineers (process, structural, mechanical, piping, electrical, instrument, and more) combine their talents to engineer the systems that make up the facility or product. Ideally, engineering should freeze all major aspects of the systems' designs, including interfaces between systems and purchasing activities, so that detailed engineering of the areas can proceed.

If a system isn't frozen, it means that someone is still thinking about it—and changing it. Take for example the heating and air conditioning (HVAC) system. Let's say that the client hasn't decided if the parts room for a large factory will be air conditioned. If the decision is later made to add air conditioning, it will increase the size and required capacity of the HVAC system. This can create knock-on effects into other systems such as electrical power or instrumentation. Those systems may be able to provide the additional loads, but if not, they will have to be expanded. That means that new equipment and more space must be added. If area engineering has started, the ripple effect will continue as the larger equipment and more ductwork cause redesign.

If the project engineers see undefined or poorly defined systems that can potentially affect their area, they should jump in and help solve the problem or escalate the issue to someone who can.

Purchase Order Formulation, Technical Requisitions, and PO Bidding

As the systems take shape, buyers and package engineers begin formulating bid packages for the purchase orders. When enough technical definition is available for a given purchase order, engineering prepares a technical requisition, containing a system description, data sheets, and specifications. It takes enormous effort to bid, evaluate, and award the hundreds of purchase orders on a large project. As a rule of thumb, the initial bids for POs should hit the street within 4 months from the start of detailed engineering, or there may be trouble. The entire project management, especially the engineering and procurement management, will watch the graph of "purchase orders placed versus time" to see if the plans are being achieved.

For EPC contracts, pricing for major items should have been negotiated during the EPC bidding, but there is still substantial purchasing and logistics to be accomplished during detailed engineering.

Manufacturing/Quality Surveillance

With purchase order award, the vendor's engineering moves into full swing. Next, materials are ordered and eventually manufacturing begins. (Manufacturing is covered later as a separate phase. This section views manufacturing from the client's and engineering and procurement teams' points of view.) The client and contractors turn their attention to quality and on-time delivery. Buyers and package engineers administer planned quality surveillance programs to ensure that products are designed, manufactured, and tested to required standards. They review the vendors' quality management plans and pick out the critical activities they want to witness or check. Experts are called in for highly technical surveillance. In addition, buyers and package engineers closely monitor schedule, cost, and, of course, the delivery of vendor information.

Vendor Data Receipt

Once the equipment is ordered, engineers must receive feedback from the vendors. They need detailed information about how the equipment will function and interface with other equipment. This whole process of receiving vendor data and integrating it into the design requires careful attention and considerable energy to keep from delaying the project. Visualize trying to collect detailed vendor data (mechanical interface drawings, electronic signal details, and more) for 400 purchase orders, awarded to vendors around the world. On large projects, a project engineer could be assigned during peak periods to coordinate vendor data gathering.

Area Engineering—Drawings, Data, and Documents for Bid Packages

As vendor information comes in, the engineering focus shifts from systems to areas. The facility must be designed in enough detail for contractors to be able to bid on the job. The better engineering houses have strategies for freezing the design that goes into the contracts while the engineering team continues to add the details needed for construction.

Drawings, data, and documentation for bids are assembled into bid packages and delivered to the bidders. For EPC contracting, there may or may not be separate contracts for construction, since responsibility for all of the engineering, procurement, and construction falls under a single prime contractor. Nevertheless, all the same engineering, equipment, and materials must be transferred to the construction sites.

The area engineering continues as the construction contracts are being bid.

Contracts Formulation, Bidding, Evaluation, and Award

An EPC contractor will generally follow a more streamlined approach for competitive bidding than was described in the "Definition" section of this chapter. This is because prices and rates were negotiated and agreed upon with subcontractors during the EPC bidding. Contract specialists compile the contracts from existing templates, and engineers add the technical specifications. Care is taken to make the contracts back-to-back, which means that all of the contractual obligations toward the client get passed to the subcontractors. Cost and schedule margins would have been added to the EPC contractor's bid to cover risks, such as

- changes due to design development,
- normal cost and schedule growth,
- anticipated risk events, and
- other unknown risks.

Competition, however, tends to limit those cost and schedule margins.

For contracting strategies such as EPCM, in which the client is more involved in contracting, there is full competition. Construction contracts are bid, evaluated, and awarded, as described in the "Definition" section.

Area Engineering—Construction Drawings, Data, and Documents and Verification of Crucial Deliverables

While the bids are out, area engineering continues, so that detailed construction drawings and specifications will be ready to deliver to the winning contractors, according to negotiated schedules in the contracts. You'll hear terms such as "AFC" (approved for construction) or "IFC" (issued for construction).

This is the time when constructability and the client's operability considerations are incorporated. Project engineers bear responsibility for coordinating both. There is much money to be saved by properly coordinating the engineering-construction interface to ensure that

- packages of drawings arrive when they are needed in the construction sequence,
- they are the right drawings, and
- they contain the correct level of detail.

Adjusting the distribution of work between the engineering team and the construction contractor's engineering team that prepares the construction work packages will capitalize on the strengths and capabilities of each. Unfortunately, adversarial relations between contractors often impede this kind of cooperation, to the detriment of all parties.

The quality of the design drawings, data, and documents is assured through an extensive quality management system that consists of interdisciplinary checks (discipline personnel checking the impact of others work on their own), design reviews, and supervisory approvals. As the design takes shape, risk assessments and safety studies are performed to mitigate any unsafe design features. Human factors are considered in the design to protect the health and enhance the efficiency of future workers.

Successful projects verify those design deliverables that are crucial to the project's success before the deliverables are sent to the construction sites or manufacturing shops. The most critical deliverables are reviewed by independent experts, while the less critical ones are checked for reasonableness by project team members. Model tests, fabrication studies, or independent analysis are other forms of verification that can be used to ensure that the engineering products are ready for use.

Equipment and Materials Deliveries to Construction Sites

Purchasing specialists on the project team track the progress of each purchase order. When equipment and materials are ready for delivery, logistical specialists arrange the movement of those products from vendors' shops to their project destinations (e.g., construction sites, other vendors' shops, warehouses, or marshalling yards). A freight-forwarding company is often contracted to transport all of a project's goods. On very large projects or in remote locations, the project may hire a company to manage the logistics (e.g., transport, store, secure customs clearances, and deliver materials and equipment).

Commissioning and Operating Procedures

A plant that processes or manufactures a product must be commissioned before production can start—it must operate safely before it can be used! This triggers events early in engineering and procurement. Engineers define systems and subsystems and then design the facility so that the means (e.g., flanges, valves, procedures) are available to be able to commission parts of a system as they become available.

Engineering prepares a commissioning plan to schedule an efficient completion of all project systems. The next task is to draft commissioning procedures that contain step-by-step checklists. The commissioning team further develops the procedures by adding practical details and operational safety measures. Operating procedures are prepared using a similar approach with the client's operating team.

The procurement team plans and arranges for on-call vendor assistance at the sites to support construction, commissioning, and start-up. They also order project spare parts and make sure system operating manuals and other commissioning and start-up documentation are delivered from suppliers on time.

Follow-up Engineering and As-Built Documentation

As the construction drawings hit the sites and are reviewed, questions will start to come back to engineering. Even though design engineers are being demobilized at this time, some stay on to answer those questions (site queries) and engineer any necessary design changes. They go to the construction sites to assist with the construction surveillance (confirmation that selected critical work products meet their requirements). As systems are completed, engineers update certain drawings to "as-built status." Preparation of as-built drawings closes the loop on engineering's contractual design responsibility. As-built drawings are compared with the construction (AFC) drawings and all approved changes, to verify that the facility or product is built as it was designed.

As commissioning begins, the engineers that remain on the project shift their focus from areas back to systems. Engineers, especially in the electrical and instrumentation disciplines, are involved in the commissioning of the systems. This occurs before the facility is opened for business and could take several months.

Contract Administration and Changes to the Contracts

Contract administrators generally move to where the action is. They stay in close contact with their management and the other party's contract administrators. When problems arise, they negotiate a settlement and process a timely change to the contract (change order). If design changes are involved, they work closely with engineers and project controls personnel to process those changes through the project's change management system, before writing the contractual change orders. This keeps technical and construction problems from growing into commercial problems, contractual claims, or even litigation. Successful projects administer their contracts in a timely, firm, consistent, and fair manner...but it takes two to tango. If one of the parties is playing games, adversarial relationships can take over.

Contracts and PO Closeouts

As the engineering and procurement activities wind down, the contracts and purchase orders are closed. Problems with outstanding deliverables, penalties, or claims are resolved. They're usually about money.

Engineering and purchasing experience is captured in many ways, including contract closeout reports, lessons-learned workshops, and PO evaluation forms.

PROJECT IMPLEMENTATION: MANUFACTURING

Many of you will work in the industry that manufactures the equipment and materials. This manufacturing is crucial to the project's success: The equipment must work as expected. The steel must be able to be welded at efficient rates. Piping materials must stand up under the operating environment, no matter how corrosive.

Some items can be bought off the shelf, but for the most part projects spend substantial amounts of money buying engineered equipment or special materials, mostly with detailed requirements specified. Each of those purchase orders is a project in itself, with its own planning and implementation that are similar to an overall project. And that project must fit in with products that the manufacturer is making for other customers.

A simplified manufacturing cycle for engineered equipment consists of at least seven main steps (see Figure 4.5):

- Planning, basis definition, and systems engineering
- Engineering
- Development
- Manufacturing and subcontracting
- Testing
- Delivery, documentation, and closeout
- Use

There are also the front-end commercial activities of submitting a bid proposal and awarding the contract. At the end is the normal feedback loop to gauge client satisfaction and improve the product and the commercial results.

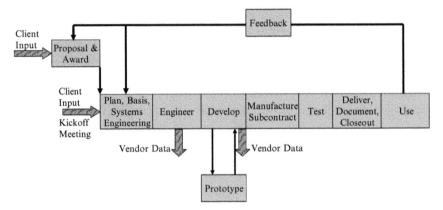

Figure 4.5   Typical manufacturing cycle for engineered equipment.

CLIENT INPUT

As with any other project, the process starts with client input—usually a bid package, negotiations, and a kickoff meeting. The quality of client basis information runs the full range from not enough to too much. The manufacturer generally struggles to get timely information on the operational environment and functional requirements for the equipment. That is information that only the client can provide. Some clients overspecify by providing boxes of general project specifications that the manufacturer must sort through to find the relevant information.

PLANNING, BASIS DEVELOPMENT, AND SYSTEMS ENGINEERING

Planning, basis development, and systems engineering also follow the familiar project pattern. The technical manager and systems engineers play key roles in planning project initiation, preparing the detailed scope of supply, compiling the detailed basis needed for design, and engineering the overall system.

ENGINEERING

Once the system is defined, design engineers fill in the details and prepare shop drawings. Often the systems engineering, engineering, and development steps are an iterative cycle until all the bugs are worked out. The manufacturer delivers interface and other design-critical information (called vendor data in Figure 4.5) to the project's engineering team. This information is needed to lay out equipment and to complete the overall project's systems and area engineering (described earlier).

DEVELOPMENT

The development activities confirm that the product is ready for commercial manufacturing. If the equipment is new and several of them are to be manufactured, it's often wise to build a prototype model, as represented in Figure 4.5. When building the prototype, drawings and specs are prepared, and all the manufacturing steps are validated. Prototyping essentially flushes out all the problems ahead of full production. Final vendor information is delivered to the project once the design is frozen for manufacturing.

MANUFACTURING AND SUBCONTRACTING

Full production begins once the design has been verified and early production units have been evaluated to confirm the manufacturing process. Production measurements monitor the quality of the product. Some components will be subcontracted.

The client will review the manufacturer's quality management plan and set up a quality surveillance plan, with witness points or hold points at critical milestones in the manufacturing process.

TESTING

This box represents the final acceptance testing, which compares user requirements from the client's input with the performance of the equipment. It can also include tests to ensure that the equipment has been properly installed.

Earlier tests during development help ensure that problems aren't pushed to the end of the job, where they are hard to detect and costly to correct. Component tests verify that individual components meet their requirements. Integration tests confirm that the components fit together and that control functions operate correctly across the component interfaces. System tests demonstrate that the product meets the system requirements formulated at the beginning of the design. They also detect errors that can't be detected in any other way. By employing all these tests, faults can be found at the earliest possible time.

TIME PRESSURE

Time pressure is a fact of life in manufacturing. Penalties imposed for late deliveries drive the manufacturer to get the equipment out the door. Competition drives companies to get new products to market. My son, who was in R&D at the time, once told me, "Around here R&D means Research and Delivery."

Projects often make the mistake of pulling the equipment out of the manu-facturing shop before it's completely finished to avoid delaying construction. Finishing equipment in the field, away from the efficient shop environment, is nearly always problematic and much more expensive.

ROLE OF PROJECT ENGINEERS

In the manufacturing segment of the business, systems engineers perform many of the project engineering duties that involve responsibility for a system from beginning to end. They coordinate the engineering disciplines and other resources, and they solve problems. Their scope includes all of the boxes in Figure 4.5, from planning to delivery. Systems engineers are responsible for managing the technical complexity of the system, its quality, and safety. Systems engineers are cost and schedule conscious, but the manufacturer's project management team will bear those primary responsibilities. For a complete discussion of a systems engineer's role, refer to *Systems Engineering: Coping with Complexity* by Stevens, Brook, Jackson, and Arnold (1998).

PROJECT IMPLEMENTATION: CONSTRUCTION

Like manufacturing, construction is a phase of the project in which plans become reality. It consists of activities needed to safely build, fabricate, and erect or install a facility. If you watch a well-managed construction site or fabrication yard, you will be amazed at how quickly the job takes shape. Structures seem to rise as if by magic. You will also be amazed at the ingenuity employed to make the worksite safe and efficient.

CONSTRUCTION SYSTEMS

But when you see a tall building rising at a rapid pace, it's not happening by accident. It comes about because the company managing the job has the neces-sary systems and procedures in place. I'll list some of those systems to give you an idea of what I mean:

- Planning systems to schedule the job (or, in a fabrication yard, all the jobs), allocate resources to the various activities, and avoid peaks in the workload that exceed capacity.
- Estimating procedures to ensure that high-confidence bids are efficiently assembled (accounting for all the risks) and approved by management.

- Project management systems that organize, direct, track, and coordinate the construction resources. An essential element is a construction manager for each job who has clear responsibility and authority to get his or her job done.
- Efficient construction processes that are continually improved to reduce the work-hours needed to fabricate, assemble, or erect the product.
- Information management systems to enable the project manager to track and control safety, quality, cost, and progress.
- Engineering systems to prepare detailed construction drawings and engineer safe and efficient construction methods (such as heavy lifts).
- Systems that put together work packages containing instructions and drawings for teams of workers to use.
- Safety plans to keep workers free of injuries and industrial illnesses.
- Procurement systems to order, inspect, transport, preserve, and store equipment and materials.
- Systems that distribute and trace which materials are used in which locations of the job, so that the materials' integrity is assured (for example, so that carbon steel and stainless steel piping materials are not mixed, creating a potentially dangerous situation when the facility is operated).
- Systems to implement changes and keep their consequences from rippling through the job.
- Quality management systems to ensure that all systems are implemented and functioning, including inspection and test plans to verify that the work products meet the specified requirements, and dimensional control procedures to make sure interfacing parts are within tolerances and fit together.

Those management systems and more are required for construction contractors to perform at a competitive level. The better contractors will use computer systems and specialized shop equipment to gain an additional competitive edge. Those contractors with minimal management systems will struggle each time they get a new job that is beyond their experience.

AREA FOCUS

For efficiency, construction work is organized by areas—the same types of areas discussed in Chapter 2. Various construction trades (structural, painting, mechanical, piping, electrical, instrumentation, and other skills) enter the areas, in a planned sequence, to do their work. Construction project engineers exercise their duties (described in Chapter 2), emphasizing coordination of surveillance and quality control for their respective areas.

Consequences of Mistakes and Changes

Any and all mistakes made in engineering and procurement will create problems during construction. Errors, omissions, and changes in the engineering drawings will lead to trouble. In the construction phase, design changes are only bad.
In the final stages of a job, changing an item can easily cost five to ten times as
much as it did to build it in the first place. That's why it's good practice to take
extraordinary efforts to find and correct engineering errors and omissions during
the first few months of construction.

Late delivery of drawings, equipment, or materials is a major source of trouble.
It forces the construction contractor to do things out of sequence, which results
in inefficiency and rework. The transition from engineering and procurement to
construction requires detailed planning and follow-up, which many projects fail
to give it. If the engineering and construction contractors haven't agreed on schedules for the drawings, documents, equipment, and materials that pass across their
interface, then hold on for a turbulent ride ahead! There is probably more finger
pointing at this hand-off (between engineering and construction) than at any other
transition on a project.

Quality problems with the equipment and materials can be equally disruptive.
Having to stop the work in an area while a vendor rebuilds a piece of equipment
can seriously delay progress. Now you can begin to see why we put contingencies
in our budgets and reserves in our schedules.

Client Intervention

The client's project team is also on the scene during construction. Their role
is to monitor the work; ensure that safety, quality, cost, and schedule targets are
being met; and manage the commercial matters. The challenge for the client's
team is to find the optimum level of intervention to fulfill their role. Too much
intervention will cost extra money and delay. Too little could allow the construction to go off the rails. The contractors' competence is a key factor in making
that call.

Focus Shifts to Systems at the End

As construction comes to an end, the focus shifts to completing systems
rather than areas, in order to avoid commissioning delays. A system, such as
the electrical power system, might run through every area of the entire facility, and those parts of the system must be integrated. Each system (or subsystem)

must be brought to a status of "mechanically complete" before being commissioned. This means that the system or subsystem is completely built, hooked up, and tested, so that it's safe and ready to be operated on a trial basis. Mechanical completion must be documented, so that everyone knows the status of all systems and subsystems. However, since completing systems takes more work-hours than completing areas, the construction group's management may resist that shift. The transition from construction to commissioning requires planning and hands-on management to avoid commissioning delays. Successful projects often mobilize a separate group to complete systems in parallel with the normal work to complete the areas.

PROJECT IMPLEMENTATION: COMMISSIONING AND START-UP

Commissioning (sometimes called *systems completion*) is the final step in preparing to start up a facility. It's generally carried out on a system-by-system basis and in a certain order. For example, much of the work done during commissioning requires electrical power. The power system is, therefore, one of the first systems to be commissioned, so that power is readily available on the site. Recall that the prerequisite for starting the commissioning of the power system is that it be mechanically complete—that the proper certificates are signed or entered electronically into a mechanical completion and commissioning database. This means that the equipment is safely installed and tested, and that the wires are pulled into place and terminated at the proper contacts. The circuits are then loop-tested to verify correctness before the high voltage is turned on. Often, a joint group, consisting of construction, commissioning, and future operating personnel, walk the system, inspecting it, and recording a punchlist of items that must be corrected by construction workers. The commissioning team uses commissioning procedures (prepared earlier by them and the engineering team) to ensure a safe, systematic approach. Sometimes the commissioning team function tests subsystems according to the procedures before eventually operating the whole system. When the electrical system is running stably, it's formally handed over to the operating team that keeps it running and maintained. The project now has power to continue commissioning the rest of the systems.

From a quality standpoint, commissioning is a significant step. It validates the design with tangible inspections and tests of the entire system under controlled operating conditions. It's the ultimate test that the design is correct. Minor deficiencies are corrected on the spot. Changes with overall systems' impacts are engineered and implemented (on a crash basis) through the project's change management process.

From an operations standpoint, commissioning allows for the gradual start-up and handover of the facility in a safe and controlled manner. Operations personnel are often assigned to the project team to fill positions on the commissioning team. The client's equipment operators are sometimes mobilized to run major equipment in order to benefit from the knowledge of project experts, at the same time testing their operating procedures before operations actually start.

Once all the essential systems are commissioned, the start-up team comes on the scene. The core start-up team often consists of operations personnel who have been part of commissioning. The leadership of the start-up team may transition from the project team to the operating team, to avoid an abrupt hand-off. When the bugs have been eliminated from the systems and the acceptance criteria are met, control passes to the operating team. The project is essentially over, except for some closeout activities, like delivering documents to operations, and commercially closing out all the remaining contracts and purchase orders (as discussed under engineering and procurement).

TRANSITIONS AND HAND-OFFS

I recently watched an Olympic relay race in which a runner made a bad hand-off and cost her team the race. The runners ran beyond the hand-off zone before the baton changed hands, resulting in disqualification. Projects are similar, except the consequences are usually time and money. You haven't fully experienced the "agony of defeat" until you've worked through a badly planned or executed transition between major project phases.

If all goes well, the client plans the project, funds it, and contracts for the work. Through the contracting process, the responsibility for the design and construction is passed to contractors and vendors. Engineering contractors pass requisitions to procurement and drawings to construction. Procurement delivers equipment and materials to the job sites. Construction is completed, and the facility is handed over to the commissioning team. Responsibility passes to the commissioning and start-up teams, and eventually back to the project team for a review of the as-built drawings to ensure that the job was constructed as it was designed. When the project is handed over to the client's operations team, they accept responsibility to maintain the integrity of the facility throughout operations, including when it is modified.

But just imagine how it can go wrong. What happens if

- the client withholds funds from the contractors and vendors and won't give them a clear signal to start, or

- the client's design basis, FEED, or bid documents are inaccurate and incomplete, or
- the contract has unclear requirements?

What happens if the cooperation between engineering and procurement is poor? In that detailed, complex, and fast-paced race, there is no time for game playing. Situations occur where engineering has allowed insufficient time in their schedule for procurement to place the purchase orders and get back the necessary design information from the vendors. Sometimes, insular company purchasing groups waste weeks bidding low-cost commodities, while engineering is clambering for interface data on those parts. Buying from a preferred supplier, without bidding, saves valuable time. Unless these two groups are managed as a unit, inefficiencies, wasted work-hours, and ultimately delays are inevitable.

What happens if

- the construction drawings are delivered late or out of sequence for the construction work, or
- the piping and structural drawings have many clashes (interference between structural members and pipes) or holds (areas on the drawings that are lacking information), or
- equipment or materials are delivered to the construction sites late or out of sequence, or
- the layout for an area is so poor that a major redesign is needed—it's back to square one?

What happens if construction finishes the systems out of sequence, and the commissioning team must wait for weeks until the right systems are mechanically completed?

What happens if the plant operates below capacity or doesn't work at all, because of faulty engineering or construction?

The answer to all these questions is, "It costs time and money." These are all major problems that can be mitigated by spending some of that time and money beforehand to plan and manage major transitions and hand-offs. It's simply good business to get the parties together, plan the transition activities, and do some team building.

BACK TO PROJECT ENGINEERING

Don't let all these potential problems burden you. At least you're now aware of them and can see them as opportunities to improve on the norm.

Now let's descend from our high-altitude reconnaissance into the world of project engineering—the world of running your leg of the race and handing the baton to the next runner. We'll use Bob Clawson's themes (plan the work and work the plan) in the next chapter while considering a case study. It illustrates the essential details of planning and carrying out a young project engineer's job, bringing to life, in a tangible way, the principles we've considered so far.

REFERENCE

Stevens, R., Brook, P. Jackson, K., and Arnold, S., *Systems Engineering* (Pearson Prentice Hall, London, 1998), pp. 5–11.

Chapter 5

Learning Project Engineering on the Job: A Case Study

You can't really understand what it's like to be a project engineer until you're there. And when you get there, the stakes may be high and the situation may be unforgiving of blunders. Fortunately, you can experience it vicariously.

One way is through the experience of your bosses and colleagues, conveyed in conversations, counseling sessions, and war stories. This experience is valuable, but it could arrive too late, since it will probably be triggered by some event on the project that's created a problem for you. And, as you know, you have to listen attentively and critically to pan out the gold and throw away what's irrelevant or exaggerated (as war stories can be).

For the same kind of benefit—letting you know what it's like to be there—but in a more controlled way, this chapter is a case study about a new project engineer, Ron, who finds himself thrown into a position of responsibility before his feet are on the ground. We'll join the project just after the contract is signed, at the outset of the project implementation phase.

CASE STUDY

The story line and characters are drawn from the example in Chapter 1. Recall that Sara Margeaux is the Project Manager for a natural gas pipeline compression station. We'll call her project the Rio Bend Project. She works for a moderate-sized company, Wizard Oilfield Contracting. Her brusque but highly competent boss is Ted Kramer. Wizard has been hired by the client, El Dinero Petroleum, to engineer, procure, and construct the project facilities.

Sara's counterpart at El Dinero is Arlo, the client's Project Manager. Dip, a crusty Operations Manager at El Dinero, wields his influence as the story unfolds.

Our main character is Ron Neuman, recently hired directly out of college by Wizard. Pay close attention to Ron's experiences as he finds his way among the more senior people in this eventful project. His boss is Edgar Baker, the Rio Bend Project's Lead Project Engineer. Ron was assigned as Compressor Module Project Engineer shortly after being hired. He's responsible for managing the compressor vendor, B&B Compressor Packaging, that will assemble all the parts of the compressor module into a complete package and deliver it to the project. Ron deals directly with the vendor's Chief Project Engineer, Jeff Collins, and occasionally with Chet, Jeff's infamous boss. Ron also deals indirectly with a subcontractor to B&B that actually manufactures the compressor assembly. Carlos is the Technical Manager at the compressor manufacturer.

Of course, all of the companies and characters you meet are fictitious, but the situations are realistic.

September 20: Sara Sends Up an SOS

It was early evening as Sara sat at her desk with the schedule in front of her, shaking her head. *There's no way I can do it! I can't make this schedule and run the engineering myself. I've got to have a Lead Project Engineer. Ron has a lot of potential, but he's young and hasn't been through this before. His compressor package needs to move without a hitch, and he will need a lot of coaching. Plus we can't let the engineering get off track. This job has an intense 12 months of engineering and procurement and that compressor is on the critical path. If I have to be worrying about that, no one will be looking ahead and planning for construction and commissioning. We'll never make the schedule and that start-up penalty will be breathing down our necks all the way.* She sketched the organization chart and scrawled the points of her argument on a pad of paper. Then she reached for the phone and speed-dialed her boss, the project's senior management sponsor. She had every reason to expect that he would still be in the office struggling to think of a way to avoid the delivery penalty for another job.

"Kramer."

"Ted, I'm going to need some more help on the Rio Bend job." Sara presented her case. As one of Ted Kramer's most reliable project managers, she had credibility with him.

When she had finished, Kramer asked some questions, thought for a few seconds, and then agreed with her proposal. "You caught me at a weak moment," he said.

I hoped I would, Sara thought. "How's the High Point job going?"

"Not so 'high' right now," he sighed. "You really don't want to know." He changed the subject. "You need to write up this organizational change, Sara, and run it through the change management system. You know you'll have to cut back somewhere else on the personnel budget to compensate for these extra costs. Fax me the Org Chart in the morning. I want to look at it. Who do you have in mind for the Lead Project Engineering job?"

"Edgar Baker is my first choice." Sara then read the other names on her list so Kramer could write them down.

"I think you're right, but it won't be easy to get Baker. I'll talk to Bill tomorrow and see if he can shake Edgar loose," Kramer said as he hung up.

Sara hung up the phone. She briefly felt pleased, but soon the smile left the corners of her lips and a wrinkle creased her forehead. *Now if Ron can just deliver that compressor on time. God, I hope we don't have problems with it. He's so green.*

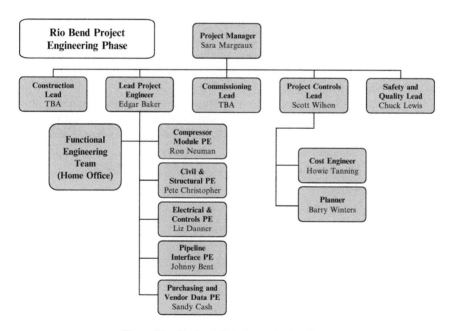

Figure 5.1 Rio Bend Project organization chart.

SEPTEMBER 21: SARA'S FAX TO KRAMER

TK:

Attached for your review and approval is my proposed Organization Chart (Figure 5.1) for the engineering phase of the Rio Bend Project. I recommend the addition of a Lead Project Engineer (Edgar Baker) to manage the engineering effort. A core team of Project Engineers and a Design Team dedicated to this project will report to the Lead Project Engineer.

Regards,

SM

SEPTEMBER 24: TROUBLE AROUND THE BEND

In a good mood, Jeff Collins returned to his office, sat in his chair, and leaned back. In the 10 years he had been Chief Project Engineer for B&B Compressor Packaging, the meeting that had just taken place was the first time that a contractor had invited him to their kickoff meeting. He liked the way Sara had stressed the importance of working as a team and especially that she had given him an office in the Project Team building. Their formal interface meetings were another plus that would help prevent surprises. Jeff didn't much care for having to work with that inexperienced package engineer, Ron Neuman, but he would have Edgar to go to if things really got bad. He and Edgar had worked on another job just a few years ago.

Jeff also liked Sara's comment on communication: "Communication within this team should be often and informal. If something happens, it should be communicated early so we can deal with it and not lose valuable time. Lastly, our communication should be candid. I don't want to have people playing games at the expense of the project's success." She had written on the board:

Communication
- Early
- Often
- Informally
- Candidly

Jeff thought he would use some of this information at the next meeting with his Project Engineers.

Then a large envelope in his overfull in-box caught his attention. *Could those be drawings from the compressor manufacturer?* he thought to himself. *They'd better be. They're running late with the interface drawings.* B&B had recently switched compressor manufacturers, at the request of the client, to avoid the quality problems they had been having. Jeff scanned the letter and found nothing out of the ordinary. It was a normal drawing transmittal letter. He slipped the CD into his computer and brought up the first sheet.

His eyes froze on a plan view of the inlet and outlet piping interface. He couldn't believe it. He paled as he switched to the next sheet to get a different view. He muttered under his breath, "I can't believe it!" *These nozzles are coming out of the bottom of the machine, instead of the top, like I ordered. This could cost us the project's schedule, the B&B profit, and maybe my job. How could they mess up like that? I even gave them the change in writing with a drawing of that detail.*

As he entered his boss's office, Jeff felt torn between anger and apprehension, and both were unwelcome and unpleasant feelings on what had started out as a pretty good Friday afternoon. Chet was sitting at his desk with his permanent scowl etched on his face. Jeff explained the problem, then took a half step backward as he took the brunt of a string of expletives that had no particular target.

"We need to tell the project about this," Jeff offered. "This could set their schedule back several months, depending on how far the compressor manufacturer has gotten."

"We're not going to tell them anything until we get to the bottom of this. One more compressor mess and the client will drop us like . . . ." Chet's words dropped off. "I'm not going to drag our dirty laundry out there until we have had some time to sort it and air it internally. Let me know as soon as you hear something else about this," Chet said as Jeff left the office.

Jeff slammed his door and slumped into his chair to collect his thoughts. The apprehension melted away, but there was enough anger left to light a fire. "I hate that guy!" he said aloud. Then he called Carlos, the Technical Manager for the compressor manufacturer, to set up a meeting.

OCTOBER 1: JEFF GETS SQUEEZED

Jeff walked into the compressor manufacturer's conference room and shook hands with Carlos and his team. "Thanks, Carlos, for arranging this meeting on short notice," Jeff said.

"No problem," said Carlos.

Being outnumbered, Jeff felt uneasy. He selected a seat near the white board and sat his briefcase on the floor. On the white board he wrote:

Objectives
- What's the progress to date?
- What will it take to get a top-nozzle compressor?

Agenda
- Status (Carlos)
- Cost and schedule impact of the changes (Carlos)
- Discussion of solutions (All)

"I think this is what we agreed on for today's meeting. Any questions?" There were none.

Carlos cleared his throat. He looked nervous. "When we got this order in the summer, we knew the job would be a foot race all the way. It's a tight schedule and the casing shop is loaded with other work. Since we knew this was for El Dinero Petroleum, the Engineering group pulled out the El Dinero design from the last job and sent that requisition to purchasing."

"Yeah," Jeff said. *Now I take the bull by the horns.* "And it took you three months to get around to telling me that you weren't following our design drawings!"

Carlos's boss, who had been sitting in the back of the room, stood up. "Look, Jeff, do we want to find a solution or do we want to shout at each other?"

Jeff boiled but held his tongue.

After a pause, Carlos continued. "El Dinero is very operations oriented, since downtime is money. How were we supposed to know they would try to save money on this job by cutting the project costs? Anyway, we placed the order for the casing right away and ordered the materials. The shop drawings are now 70% completed. The first release went out last week and included the interface drawings that you received."

Jeff knew that the casing order had locked in the schedule, but he pushed back anyway, appealing to their sense of fairness. "The problem is that we—my company and yours—bid the top-nozzle design that the client requested. Now we're giving them a bottom-nozzle design. What's it going to take to get this thing straightened out?"

Carlos looked at his boss as he spoke. "It's going to cost us $1.2 million to redo the engineering, purchasing, and manufacturing we've already done. The casing is 'long lead' and will add 6 months to the schedule. We have to pay the casing manufacturer a premium to get back into the queue."

Jeff absorbed this incredible news. "Look, we signed a contract with you guys to deliver a top-nozzle compressor and that's what we have to deliver to the client."

"Maybe you'd better shop around," Carlos said, knowing full well that the client had specified his company as the designated subsupplier. "That's the best we can do—$1.2 million and 6 months. Those top-nozzle drawings you sent us were never part of the contract. By the time we got them, we had already placed the casing order. How was I to know the El Dinero design had changed?"

Jeff got into his car and dialed Chet's number from the parking lot. This time he wasn't going to suffer another face-to-face encounter with his boss, at least, on this subject. "Chet, this is Jeff. It's worse than I thought. They placed the casing order 3 months ago based on the standard El Dinero design. They're squeezing us for $1.2 million and 6 months' delay. Carlos's boss was in the room, so I know it's the company position." Jeff summarized the rest of the meeting.

Chet thought for a moment. "Don't tell Wizard Oilfield and don't send them the interface drawings. Just continue like nothing's wrong. Somehow, we have to pass this cost on to the contractor or the client."

"But the Rio Bend Project needs to know this!" Jeff protested, but the line had gone dead.

OCTOBER 1: EDGAR JUMP-STARTS THE ENGINEERING

As Edgar Baker stood, the conference room became silent. He was a tall man with angular features that give him a commanding presence. As he spoke, however, Ron noticed a subtle smile that reflected from his eyes and the corners of his mouth. *Edgar seems like someone I could like and respect*, Ron thought.

Edgar wrote on the board.

> Our goal is to complete the engineering and procurement scope of work by Labor Day of next year.

"I thought you said we had a whole year? That's only 11 months!" someone in the back said, and everyone laughed.

"Glad to see the team is in good spirits," Edgar said as he looked over his shoulder, and then continued to write and talk.

> Today's objective is to plan the work.

"There are several things I would like to cover," Edgar said as he passed out the agenda.

Agenda
- Team Introductions and Team Process
- Establishing Design Basis; Check Client's Input Data
- PSB
  - Planning (scope of work and deliverables)
  - Scheduling (network diagram, level 4 & 5 schedules and 30-day plan)
  - Budgeting (engineering costs)
- Interfaces and Summary of Responsibilities
- Meetings
- Problems

"I'll start the introductions, and then we'll go around the table. I'm Edgar Baker, Lead Project Engineer for the Rio Bend engineering." He went on to describe his tenure with the company and his recent positions. The others did the same. Ron's nervousness escalated with each introduction, but he resolved not to let his lack of experience erode his confidence. When it was his turn, Ron said, "I'm Ron Neuman and I'm new around here. My job is to get the compressor module to site by next Labor Day with no punchlist items and no excuses." He looked to the person beside him and passed the floor over with a nod.

Edgar passed out a sheet of paper. "This is the way I would like our team to function," he said. "The first point is probably the most important. We have to keep accurate information flowing, both amongst ourselves and with other organizations. In other words: *no surprises and no games.* We also have to keep our credibility high. Once it's lost, it's impossible to regain."

Rio Bend Project Engineering Team Process

- Communicate:
  - Early
  - Often
  - Informally
  - Candidly
- Keep our credibility high.
- Use benchmark data to plan, schedule, and budget.
- Be involved—hands-on project engineering and technical leadership.
- Don't say, "No." Say, "Let me work on a way to help."
- Develop effective relationships (project, client, subsuppliers, and management).
- Expand your role.

Ron listened carefully and took notes in the margin. The last item, expand your role, had never occurred to him. Feeling more at ease than he'd expected to be at this point, he asked Edgar what it meant.

Edgar answered, "It's easiest for you to just handle your own job. But what I'm asking from each of you is to take on more responsibility. If there's a problem, take a leadership role in finding the solution. If there's an issue that cuts across other areas or even other companies, pursue that issue until everyone involved agrees on a solution. If some task needs coordination, step up and volunteer to do it."

Ron nodded his head. It made sense, and it sounded like a doable way to get ahead in Wizard.

"Are there any more questions?" Edgar asked. "If not, let's go to the second agenda item. Willie, where do we stand on checking the input data?"

Willie, the Process Discipline Lead, stood up. "As you know, according to the contract we must complete our review of the client's input data and design basis by November 15th. We had our first meeting of the Discipline Leads last week, and I expect their responses by mid-October. We'll then review the results with the Project Engineers."

"I'd like to attend that Project Engineers' meeting and then review the draft letter to the client at least two weeks before it goes to Sara for signature. Have we found any serious deviations yet?"

"No, Edgar, it seems pretty straightforward so far," Willie said.

"So much for the client's data, how's the progress on building our own detailed design basis? Have the Project Engineers been involved?"

"Not yet," said Willie.

"Pete, as Civil & Structural PE, your work is on the critical path. Those construction guys want to start the earthwork as soon as they can. You're deputized to lead the Project Engineers in getting the design basis together. See me after the meeting and I'll explain what needs to be done."

"Sure," said Pete.

"Ron, I want you to get those compressor package interface drawings by the end of next week. Mechanical tells me we can't finish the site layout without them. We can't start digging if we don't know where everything goes."

"Yes, Sir," said Ron. *How am I going to do that? I'd better call Jeff this afternoon.*

"Next let's go to PSB: planning, scheduling, and budgeting," Edgar said. "Probably the most important thing we have to do is get the scope of work and the deliverables right. I sent the scope of work from our bid proposal to each of you with the meeting notice. I would like the Discipline Leads and Project Engineers to update it with comments and get it back to me by next Friday.

"I've asked Barry, our planner, to take the lead on planning and scheduling. Discipline Leads should get your list of activities to Barry by next Friday. Once Barry gets your list, he'll come around and help you work up your part of the level-four, i.e., activities schedule. Also, give him your estimate of the hours for each activity. I would like our level-four schedule completed by November first.

"On that same day give Barry your list of engineering deliverables—plans, engineering drawings, documents, electronic registers containing engineering data, and other items produced during the design. Barry will then work with you to pull together the level-five schedule of all the engineering drawings, documents and electronic registers."

"Wow!" said Walter, the experienced Mechanical Lead. "I've never seen anything like this around here before. This is more planning than we've ever done on a job like this. It'll cost us more to plan the work than it will to do it. And, by the way, I don't need anyone's help to draw up a schedule."

"Bear with me on this, Walter. Let's talk about it offline and I'll tell you what we're trying to do. I think this is a good time for a break. Be back in 15 minutes."

OCTOBER 1: CHET TAPS THE "GOOD OLE BOY" NETWORK

Jeff sat in his car, staring at his cell phone. He shook his head from side to side. "I wonder if that S.O.B. heard what I said?"

Chet could hear Jeff's protest as he hung up the phone, but he was already making his next move. *Now who's the best operations guy at El Dinero to call?* "Hello Dip, is that you? Hey buddy, this is Chet. How's it goin'?"

"Well, I'm sitting here with grease all over my face and coveralls from one of your #@! *&#! compressors."

Bad timing, thought Chet. "Want me to call back later?"

"No. I want to talk to you right now while this is fresh on my mind." Dip went on and on about the problem he was having, with Chet contributing an occasional "Uh-huh."

"I think I know who can help y'all, Dip. We have an expert on that model over here. I'll send him over this afternoon. I never heard of that problem before," Chet lied. "I'm certainly glad I called to check up on what's going on."

"Go ahead and send that expert over this afternoon, but he'd better not cost us anything," Dip said.

"Say, Dip, I heard something the other day you ought to know. Have you heard about the Rio Bend Project?"

"Yep."

"I heard that their compressor design has the nozzles coming out the top of the casing. I know how you operations guys hate that because you have to take out all that overhead piping every time you service or repair it. I just thought you might want to know that, ole buddy."

"Thanks for letting me know that," Dip said, with a serious tone in his voice.

"OK. I'll be seeing you, and I'll send that serviceman over today," Chet said.

October 1: Planning Meeting Continues— The Compressor Is a Problem

Edgar called the meeting back to order after the break. Ron was glad to see that Walter was back to being his calm, reasonable self. Edgar was Ron's boss, but Walter was his mentor and had helped him with the compressor package. Ron didn't want to get in the middle of a struggle between them. *They must have talked during the break*, Ron thought.

Edgar spoke. "Let me summarize scheduling. With all of your help, Barry is going to put together a planning network for the whole job—engineering, procurement, construction, and so forth. Management has asked us to prepare more detailed level-four and -five schedules than normal. The company has had schedule problems recently and is facing some big penalties."

"Yeah, that High Point is jabbing us all in the backside," someone said.

Everyone laughed and even Edgar, who had been the stone-serious company man, had to smile. "That's one way to put it but, whatever the reason, we all need to plan the work and work the plan. And I want the Project Engineers to squawk when we're not on schedule." Edgar continued, "Barry will also build a rolling 30-day plan and publish it every Friday at our weekly meeting."

Feeling ignorant, Ron squirmed but decided he needed to know more than he needed to appear to know. "What's a rolling plan?"

"It's a bar chart that covers the schedule of everyone's most important activities for the next 30 days," replied Barry. "I'll update it before the Friday meetings and show where we stand on each activity by drawing a frontline. We will do this at least through Christmas."

"Ron, you will have to work with Barry to get the vendor's schedule for the compressor package and integrate it into the overall schedules. It's on the critical path of engineering, procurement, and construction," Edgar said.

That's the second time he's said that, Ron noted. *Must be important. Sounds like if I screw up, I screw up everybody downstream.*

"In the interest of time, let's move on to budgeting," Edgar said. "Howie has been assigned as our Cost Engineer and will lead the charge on the budget."

"I'd like all of you to give me your team's work-power estimates and costs," Howie said. "When you Discipline Leads submit your schedules to Barry, figure how many people each task will take and any other costs your team will have. I'll take Barry's data and send you a spreadsheet to fill in. I'll also work with the Project Engineers and the Procurement Department to estimate the costs of the subcontracts and purchase orders. It should be painless—just like oral surgery."

No one got his joke. They were too busy adding up the work that PSB was going to take and all the things that were due by the next Friday.

Edgar broke the silence. "Next are interfaces. We don't have much on that today but I want Johnny to mark up the company's standard interface procedure for the Rio Bend conditions and get it back to me by Thursday."

"OK." Johnny Bent was the Project Engineer for the pipeline tie-ins and had done this several times before. He knew what he was doing.

Ron saw Johnny's confidence and made a note to get his help with the compressor package interfaces.

Edgar continued, "I will manage the interfaces with the client on engineering matters. The first issue will be the review of the client's design basis that I mentioned earlier. I want to be informed of any problems or issues as soon as they come up.

"I'll also shepherd the project interfaces with procurement and construction. Sandy, in addition to your job as 'Vendor data sheepdog' ..."

Sandy smiled at Edgar's turn of phrase.

Edgar continued, "I want you to help me chase the purchasing deliverables. We have a lot of requisitions and data sheets that must be prepared and delivered on time. Also there is a standard 14-week chain of purchasing events for each purchase order that I want you to help me follow. We will want to organize a team-building session with the purchasing group in the next two weeks.

"I'll summarize all the responsibilities we discussed today in a responsibility matrix and attach it to the minutes," Edgar said as he brought the discussion about interfaces and summary of responsibilities to a close. He pressed ahead to complete the agenda. "Our weekly meetings will be scheduled every Friday afternoon."

Everyone answered, in unison, with a loud assortment of "Ahs" and "Boos."

"OK, Friday morning at 8:30," Edgar said.

That was better.

"I have my staff meetings with Sara on Monday morning at 8:00 and I need fresh information for that. Now for the last agenda item, does anyone have any problem areas we should discuss as a group?"

There was silence. It had been a long meeting and, besides, everyone already had plenty to do.

Then Ron took a deep breath and spoke up. "I'm having a problem with B&B, the compressor package vendor. When I try to get the interface drawings and their schedule, Jeff, their Chief Project Engineer, says that he doesn't work for us since the purchase order hasn't yet been assigned to us by the client. I went to the Purchasing Group and they said that Sara was going to sign that agreement on Monday afternoon."

"That's what I hear," Edgar said.

"Well, if I were Sara, I wouldn't sign it. We don't know anything about the package or the schedule or anything else. And I don't like their attitude. They would rather work directly for El Dinero than for us. I think they're hiding something," said Ron.

"I hear you," said Edgar, after a thoughtful pause. "Keep pressing on them for the interface drawings."

OCTOBER 4: SARA DIGS IN HER HEELS

Sara stared across the table at Chet, the B&B Compressor Packaging Manager, and Arlo, the El Dinero Petroleum Project Manager. "What do you mean, I'm being obstinate?" she raised her voice. "Arlo, your project team ordered this compressor package earlier this year before we were even in the picture. There was one paragraph in the contract directing us to accept assignment after contract award. That and the purchase order are all we got. We haven't received the interface drawings, and we haven't received the schedule. Nothing!"

"Look, Sara," said Arlo. "This is a routine matter. You take over the purchase order and run it according to the contract."

"And who stands behind the lump sum price?" Sara asked.

"Your company does, of course."

"My people are telling me this one isn't routine at all." She looked at Chet. "Before I can sign that agreement, we need to get the interface drawings and the B&B integrated schedule, complete with the compressor manufacturer's schedule."

"We still don't have that from the compressor manufacturer but we're trying," Chet said contritely.

"We're getting nowhere," said Arlo. "Sara, have your Project Engineer set up a technical meeting this week to sort out the details you need. We have to get to the bottom of this. Send me the minutes before Friday," Arlo said as he got up to leave.

"We'll call the meeting for 2:00 PM on Wednesday. Have the compressor manufacturer there," she said to Chet.

No way, thought Chet.

Ron and Jeff sat in silence on the same side of a conference table littered with drawings, specifications, and schedules. It was 6:00 PM, and they had been meeting for four hours.

This guy, Ron, really asks some good questions, thought Jeff. *He's young but he's sharp. And he seems like a straight shooter.*

Ron spoke. "So, one more time, you're telling me that there is no problem with the design and no problem with delivering the machine by Labor Day. However, we're two months late receiving the manufacturer's interface drawings. It just doesn't add up. How do we all make up the two months?"

Jeff avoided Ron's stare and said, "I wish I were able to do better."

"All right, that's the way I'll write the minutes, but I can't support assignment of the PO until we receive something tangible that indicates your subsupplier is on board," Ron said with exasperation.

This wasn't what he'd imagined engineering would be like when he was in college.

"Hello, Sara? This is Arlo. Say, I got your Fax with the minutes of the Wednesday technical meeting. This is just incredible! The compressor packaging company says the package is on schedule, but your guy doesn't believe them."

"I assume you noticed that there wasn't anyone there from the compressor manufacturer," said Sara. "That was part of the deal as far as I was concerned."

"I was told that they couldn't make it on such short notice. Sara, I think it's time we had a design review to get everything out on the table. We can resolve this compressor issue, and at the same time Edgar can report on your review of our design basis. I can bring our operations people and have their questions answered, as well. If that meeting doesn't clear things up, I'm going to take this assignment problem up with the El Dinero management."

"That shouldn't be necessary," said Sara. "All we need is a little information and we're ready to sign." What Sara didn't need was a problem with the El Dinero senior management for Kramer to deal with.

"Set up the design review for Wednesday the 13th. I'll send our list of questions over by noon on Monday," Arlo said, as he hung up.

OCTOBER 13: THE DESIGN REVIEW HITS THE FAN

Edgar had just finished summing up his preliminary conclusions on the client's contractual design basis. The pipeline gas specification was incomplete, but Johnny Bent, the Project Engineer for the pipeline tie-in, had agreed to call a meeting to resolve the matter.

Walter replaced Edgar beside the projector and began covering the compressor design and the site layout drawings. "We were able to implement your ideas, Arlo. It looks like we can save $400,000 to $500,000 by using a compressor with the inlet and outlet nozzles coming from the top of the compressor casing. It will make the foundation less complicated and reduce the structural costs."

"That's great," said Arlo, his face beaming.

"Well, I don't like it one damn bit," said a gruff voice from the chairs along the back wall.

"Let's take a break," said Arlo. "Sara, do you have an office where Dip and I can talk?"

"Sure, use the table in my office. It's two doors down the hall to the left. We'll adjourn until you return."

"What do you mean you don't like it, Dip? Don't you like to save the company money?"

"Your stupid cost-saving ideas will cost us more in downtime than you ever dreamed. Even routine maintenance of the compressor with top nozzles will take two and a half days to remove, replace, and test that high-pressure gas piping. That's over $1 million in lost gas sales right there. I won't have it! The standard El Dinero design calls for bottom nozzles and that's what it's going to be—period."

"Dip, you know what this will do to the schedule. If we have to change the compressor design now, there's no telling how much time we'll lose and the schedule is already tight in the base case."

"Arlo, I don't care about your schedule. That's your problem." Dip got up from the table and left the building.

Sara thought Arlo looked like he had just come from a traffic accident. "Sara, I don't think we need to go any further with the design review until we get this nozzle issue resolved. Would you have your people price the change to bottom nozzles so we can study our options? Submit a Change Request Form, complete with drawings, a bill of materials, vendor prices, and the schedule impact. I need it by October 28th."

"OK, Arlo, but that's only two weeks. It'll be tough to turn this around by then without the cooperation of your compressor packaging vendor. You'd better talk to Chet."

"I'll call him," said Arlo, as he got up to leave.

OCTOBER 13: CHANGES COST BIG TIME

Chet could hardly contain himself as he hung up the phone. Dip had come through. He swung his chair around to his computer and fired off a note to Jeff.

> Jeff: I just got a call from El Dinero's Rio Bend Project Manager. We will be receiving a request from Wizard Oilfield to price a change for moving the compressor nozzles from top to bottom. I want you to use the numbers you got from the compressor manufacturer and add another $600,000 for our costs. That's $1.8 million for the change, and 6 months' schedule delay. You will have to fill in the details. Go ahead and send them the bottom-nozzle interface drawings.—Chet

It only took Jeff 10 minutes to show up in Chet's office. He waved a copy of the email. "What do you mean by this? The compressor already has bottom nozzles. We can't charge a client 2 million for nothing. I won't be a part of this."

"You will, if you want to keep your job," said Chet.

Their eyes locked. Jeff returned to his office and slammed the door.

OCTOBER 28: GOOD WORK BUT BAD RESULTS

Ron, with three copies of the change request tucked under his arm, walked with Edgar down the hall to Sara's office.

"That's a nice package, Ron. It looks like you got input from all the right people."

"Frankly, I was stunned at the cost," said Ron. "When I added it all up and it came to $4 million and 6 months' delay, I couldn't believe it. The biggest number was from B&B. But we also had to repeat a lot of design work. That top-nozzle design doesn't look like such a good idea now."

"Changes always cost more than you think, since it's impossible to see all the knock-on effects. However, this one is a whopper."

They entered Sara's office and sat at her table. She stepped from behind her desk to join them.

"Here is the change request package, Sara," Edgar said. "The bottom line is $4 million and 6 months' delay."

"It couldn't happen to a nicer bunch of guys," said Sara.

"Ron did a good job pulling this together," said Edgar.

"Give me a moment to read through this," Sara said. "I may have some questions. Ron, did you get the interface drawings from the compressor manufacturer?"

"Yes, finally," said Ron. "They're on the CD-ROM."

"Did you follow up on any of these vendor quotation numbers? I don't see any cost breakdowns from the compressor manufacturer and theirs is a big number."

"No, I didn't have a chance," said Ron. "The package came in late yesterday afternoon. Some new guy delivered it. He didn't know anything about it. I tried to call Jeff but he wasn't in his office."

"OK, I'll sign it, since it's due today," said Sara, "but follow up with B&B to get the backup documentation. Ron, I want you to deliver this to Arlo, yourself, in case he has some questions."

NOVEMBER 3: CRUNCH TIME

Sara, Edgar, and Ron sat across the El Dinero board room conference table from Arlo and what looked like half of his team. They were clearly outnumbered.

Arlo opened the meeting. "Sara, we have carefully reviewed your change order request," said Arlo. "According to our data, the compressor changes shouldn't cost that much. We have also met with our operational people and determined that page 79 of the compressor specification requires your company to engineer 'operability' into the design. The top-nozzle design is clearly not operable.

"We also reject your notion of a 6-month delay. The project has just started and there is plenty of time to recover the schedule. I intend to impose the $4000 per day contractual penalty if the project's start-up is late."

Sara did the math on her tablet. A $720,000 penalty plus $1.5 million to reengineer the design was over four times the High Point Project's loss. *I don't like to set records like that*, she thought to herself. *Wizard Oilfield would probably be able to avoid the actual additional compressor costs but would have to take the client to arbitration court. Suing the client is seldom a good course of action.*

As he looked across the table at Arlo, Edgar slid a document in front of Sara. She took the hint and began to speak.

"These are the minutes of meeting from our contractual negotiations. They summarize your Engineering Manager's suggestion to pursue a top-nozzle compressor to save costs. How can you now claim that we should bear the burden to switch back to your normal bottom-nozzle design, based on some obscure clause buried in a spec? We will continue to work on the basis of the contract, until we receive a proper change order."

"Well, here it is," said Arlo, as he sailed the letter across the table. "This letter directs you to change to a bottom-nozzle design under Article Seven of the contract. The cost and schedule consequences will be negotiated. The Engineering Manager didn't have the financial authority to approve that change to our standard design."

Edgar tried to make eye contact with his counterpart, the Engineering Manager, but he was staring down at the papers in front of him.

"You did what?" shouted Arlo's boss. "Arlo, do you know what your decision could cost El Dinero, if Wizard Oilfield stops working or is late with start-up?" He pulled a pad of paper from his drawer and began calculating. The lead of his wooden pencil broke, and he snatched another from the homemade "World's Best Dad" pencil holder beside him. When he had finished, he held the pad in both hands, so Arlo could read it from where he stood across the desk. "Go fix it," he said. The number at the bottom of the pad, underlined three times, was $40 million.

NOVEMBER 8: JEFF SHOWS UP

It was already dark outside as Ron turned out his office light and closed the door. Down the hall, Jeff's door was open and the light was on.

"Where have you been, Jeff? I've been looking all over for you."

"Oh, I've been having a little disagreement with Chet," Jeff said. "I really don't want to go into all the details, but I'm going to be leaving B&B. I can't work for someone like Chet," Jeff said as he continued to pack the single box in the center of his desk. "I won't need this CD-ROM of the drawings for the compressor. I'll give it to you."

Ron said, "I think I already have it. Is that the one that came with the change order request?"

"Yeah, I burned the CD," said Jeff. "It was the last thing I did for the Rio Bend job...almost. Chet and the other guys made up the cost and schedule numbers for the change request."

Now Jeff was booting up the computer. He popped in the CD and scrolled through several drawings. He stopped on one that had been produced by the compressor manufacturer.

"I better go," said Ron.

"You need to see this drawing from the compressor manufacturer." Jeff said with an edge in his voice. "Look at this note on the drawing."

STANDARD EL DINERO DESIGN

"Here, sit down in front of the screen. What's the date on the drawing?" said Jeff.

"It looks like July 22 of this year...." Ron pointed to the screen but trailed into speechlessness.

"They're still working to these drawings, Ron. The standard El Dinero design has bottom nozzles," Jeff said. "There is no change. There is no delay. There is no extra cost for the compressor. Both B&B and Wizard will have to redo some engineering, but that can be made up without a penalty. El Dinero will surely pay that and might even offer an incentive to recover the schedule."

"Thanks Jeff. Hope to see you around."

"You're welcome, amigo."

Ron was in Edgar's office first thing in the morning. Both of them smiled as Ron related his conversation with Jeff.

"If we call a compressor interface meeting with both B&B and the client represented we can get the whole situation resolved and documented," said Ron.

"I agree with your strategy, but you need to move fast," said Edgar. "Try to arrange the meeting tomorrow. I can't make it, but I'll call my counterpart over at El Dinero and ask him to attend. We will want some client clout there to make sure things get resolved. It should be an interesting meeting."

The compressor interface meeting started as usual with Ron asking for the status of the outstanding issues. Chet attended since Jeff had left the project and this was an important meeting. Edgar's El Dinero counterpart, the Engineering Manager, was seated at the head of the table.

"As you know, Chet, we received the interface drawings as part of the change order package, so we can check that off the action list," said Ron.

"You did?" asked Chet.

"Yeah, they were on the CD-ROM." Ron switched the projector on and brought up one of the drawings on the screen. "I took the liberty of calling the compressor manufacturer and verified that these are the drawings that they are using in the shop. Since they prepared these drawings back in July, it appears that the whole 'top nozzle, bottom nozzle' issue was a misunderstanding."

Chet offered no explanation, so Ron continued.

"I've prepared this interface agreement with the list of vendor data and deadlines that Jeff and I had worked up. I would like you to sign it today, Chet, so we can reschedule our engineering. This information is critical to planning the project's recovery schedule."

Chet nodded in agreement.

"I would also like to sign it to confirm El Dinero's agreement," said El Dinero's Engineering Manager. "That should expedite the approval process."

November 16: The Negotiation

Sara shook hands with Ken, Arlo's boss. "Nice to see you again, Ken. Hello, Arlo." She took her place on the opposite side of the table. Edgar greeted Ken and Arlo and sat down beside Sara.

"Sara, your people really did some excellent detective work to get to the bottom of this change order," said Arlo. "I expect we are now back on the original budget and schedule."

"Maybe some people are, but we're not. We've been working to El Dinero's design basis ever since we signed the contract two months ago, and it had top nozzles. Edgar and his team have estimated the engineering and procurement hours it will take to recover."

She slid two packages of paper across the table to Ken and Arlo. They studied the summary page.

"As you can see, $1.5 million is the cost, which includes the B&B costs. We can't go any lower than that, unless you want to make this a reimbursable rather than a lump-sum job," said Sara. "The schedule delay is 2 months. We have to repeat the early engineering and purchasing, as I said, and that's on the critical path. It's not the type of work that we can make up with extra people."

She, Arlo, and Edgar argued for nearly half an hour without making any progress. Finally Sara said, "Ken, I don't think we are getting anywhere. There is just too much risk for us to accept the original start-up date, after suffering a 2-month delay. That would be like asking for a penalty. This is one time that you can't whip the horses harder. You are going to have to change the finish date."

"What if we extend the penalty date by two months, but give Wizard an incentive payment to meet the original schedule?" asked Ken. "Then you can try to find a faster route for your horses to take."

Sara considered it for a moment. "That's possible, depending on the amount of the bonus, but we will have to have more cooperation from your vendor, B&B."

"You will have to accept assignment of the B&B purchase order or the deal is off," said Ken.

"Then you will have to get Chet off the job."

"That's already been done! I talked to the President of B&B last week."

NOVEMBER 22: A PAUSE TO ENJOY AND PONDER

The Monday morning sun shone through Ron's office window as he set his briefcase on the floor and adjusted the blinds. He still felt pleased at the way things had turned out last week. On Friday, as Ron was leaving work, Edgar asked him to think about what he had learned from the nozzle problem. Ron had gone backpacking with a friend over the weekend, so he had time for his thoughts to gel in the crisp mountain air of West Texas. As he sat at his computer, capturing those thoughts, Edgar walked in, closed the door, and sat in Ron's only guest chair.

"Ron, Sara asked me to tell you that we plan to keep you on the project through construction, commissioning, and start-up. You've proven yourself by the way you've handled the nozzle issue with B&B. A lot of our success depended on the way you asked questions, pushed to get the interface drawings, and, at the same time, were able to build an effective working relationship with Jeff under difficult circumstances. Had Jeff not trusted you, we might not have discovered Chet's deception."

"Jeff has integrity. As a matter of fact, I've thought about your question and was just starting to write my thoughts when you came in," said Ron.

"You mean you didn't work on this while you were backpacking?" said Edgar.

"Yeah, I did, but the coyote ate my homework," said Ron.

"What did you come up with?" asked Edgar.

"Well, I'm sure you'll be proud of me for organizing my points under your two favorite categories: Plan the Work and Work the Plan."

Edgar smiled. "You're learning."

"I'm not brown-nosing when I say that your kickoff meeting with the Project Engineers and Engineering Discipline Leads will be my model for planning the work. They didn't teach us about PSB (planning, scheduling, and budgeting) in school. For instance, I hadn't realized that a detailed scope of work is so important."

"Yeah," said Edgar, "It's a complete description of what we do on the contract. We use it to price the job and record what we negotiate with the client when we sign the contract. In addition, it's the roadmap for the planning in PSB."

"The other part of planning that caught my attention was building a complete design basis, verifying it, and then closing the loop with the client. I think that was what first put us onto the problem with the compressor. When we couldn't get a straight answer, I had to start asking more questions."

"I think you're right," said Edgar.

"And, as for scheduling and budgeting, I still have a lot to learn," Ron continued. "But having those cost and schedule engineers on the team gave me direct insight into the details of building our level-four and -five engineering schedule. I'll be able to do it quicker the next time."

"That's good, because the revised schedule and budget are due from you today," said Edgar. "What about working the plan?"

Ron thought for a moment then said, "One thing I learned about project engineering is that even though I'm in a leadership role, I still have to know the technical business. There are a lot of technical judgments that have to be made on a day-to-day basis. If I didn't have the engineering background, it would have been difficult to make the right decisions." Ron continued, "The experience in solving problems and commercial issues is harder to quantify. I want to say that it's the awareness that every problem will have a commercial impact. From now on, I'll know to take that into account.

"Also, as you said when I first arrived on the project, interfaces need to be managed. I plan to hang your advice on my wall. That's how we finally resolved the nozzle problem," said Ron.

"You'll also want to learn more about how to deal with risks," said Edgar. "At the beginning of the job and at the beginning of each new phase, it's important to identify risks and to plan measures to mitigate them. Also when a problem pops up, start considering how you can reduce the risk it will cause. For example, after I heard what El Dinero's operations guy said at the design review, I started planning the design changes for bottom nozzles before the change was even proposed. I didn't want us to lose any more time than we had to if the change was made...and I had a good idea that it would be. It's always a matter of judgment. In this case it paid off. Because of that plan, we may be able to earn that schedule bonus if we're lucky.

"You'll certainly learn a lot more about controlling safety, quality, cost, and schedule when we get into the construction," Edgar continued. "Those four project engineering duties are probably the most important when it comes to working the plan. On the other hand, they will be more straightforward than what you have just been through. We have company systems to help us monitor and control those four objectives, but believe me, it won't be easy."

"Probably the lesson I'll remember the longest," Ron said, "is that some people don't always tell the truth. At least they don't necessarily volunteer information that is to their advantage to withhold."

Edgar countered, "Let me say it a little differently, by using Sara's philosophy on communication. Communicate early, often, informally, and candidly. It's a part of the teamwork I mentioned earlier. If someone can't play by those rules, they will have to be dealt with. That's particularly true if it's a matter of business ethics or the law. It's too bad you had to run into that on your first job out of school as project engineer."

"I see what you mean," Ron said.

"That's the difference between school and work," said Edgar. "Experience gives the test first and the lessons afterwards."

As the conversation wound down there was a loud rap on Ron's door. Edgar nodded toward the door, "Sounds urgent."

"Come in," said Ron.

Liz Danner, the Electrical and Control Project Engineer, burst into the room. "Oh. Hi, Edgar, I didn't know you were here." She turned to Ron. "Ron, you know that change we had to make on the compressor package?"

"Yeah, we were just talking about it."

"Well, piping and structural have just routed their stuff right through my control room. You need to get involved!" she said.

"Here we go again," said Ron, smiling at Edgar as he walked out the door.

REFLECTION ON THE CASE STUDY

This story paints a picture of what a project is like. By observing the characters, their thoughts, and their actions, you get the feeling that these are ordinary people trying to get their jobs done. Project people have a strong sense of teamwork and a high tolerance for helping the new engineers get their feet on the ground.

But what about Chet? Fortunately, there aren't too many like him. He's in there for a reason, which you'll discover in Chapter 7. I don't want you to get the impression that all vendors do the kind of unethical things that Chet did. He could just as easily have been on the client's team or the contractor's team. Being a vendor just happened to fit this story. Most vendors I know have integrity and try to turn out quality equipment or materials at the price they quoted.

Hopefully, you now have some insight into how to plan the work and work the plan. You've probably gathered that it's a continuous process. You plan something, and then you follow through and do it. Along the way you work as part of a team to control the safety, quality, cost, schedule, and environmental objectives. There isn't any magic about it. You manage interfaces, adapt to changes and carefully work through their consequences. You solve problems as they come up, and you take care of the money (commercial issues). And, simultaneously, you're probably planning or working several other issues.

We could have gone through more examples of the project engineer's duties, especially quality, cost, or schedule control. However, most companies have their own systems for doing this. With your company's systems, your own wits, and the advice in Chapter 2, you'll be able to handle these duties.

One of the most unexpected things that Ron discovered was that *how you approach the job* can be nearly as important to your success as *what you get done*. We'll consider that in the next chapter.

Chapter 6

Skills That Can Get You Ahead

If a man is called to be a streetsweeper, he should sweep streets even as Michelangelo painted, or Beethoven composed music, or Shakespeare wrote poetry. He should sweep streets so well that all the hosts of heaven and earth will pause to say, here lived a great streetsweeper who did his job well.

Martin Luther King, Jr.

PERSPECTIVE ON GETTING AHEAD

WHAT DOES IT MEAN TO GET AHEAD?

What does it mean to get ahead? The answer to that question lies somewhere in your aspirations, which for each of you are different. For some, it means promotions and high positions. For others, it's power or money. It could mean the satisfaction of accomplishment—of doing a good job and perhaps being recognized or rewarded for it. For still others, it means working as a member of a project team and helping to make that team successful. Some of you may have not even thought about it. For all of you, it hopefully includes an element of interest, enjoyment, or satisfaction that makes you want to come to work each day.

It's your aspirations and expectations that define what "getting ahead" means for you. If those expectations are compelling, they'll eventually help you succeed.

WHAT DOES IT TAKE TO GET AHEAD?

As Doctor King eloquently implies in the opening quotation, getting ahead has something to do with how you approach the job and how well you do it—no matter what that job is.

This chapter is built around the advice of an insightful senior executive (Petkovic, 2003). Her perspective on getting ahead, which she calls Business World 101, has three elements:

- Competence
- Office politics
- Social skills

Since these apply to all phases of one's career, let's narrow the discussion by concentrating on skills that apply to newly hired engineers as they progress toward becoming project engineers in the first five or so years of their careers. We'll also make it relevant for all levels of aspiration—from those who want to enjoy rewarding technical careers, to the most ambitious who want to become president of their companies.

Below, Petkovic's three elements of getting ahead are covered in their order of importance for you. Technical competence is by far the most important in the early years of your career. However, don't be put off by the last two. They exist and can be made to work in your favor. Politics are there in your workplace and must be recognized and dealt with. Social skills can become an extra benefit under the right circumstances.

COMPETENCE

First and foremost, competence involves your technical skills and work ethic— you develop your technical skills, and you work hard on the right objectives. Competence includes your personal efficiency and effectiveness, which has to do with how well you do your job and how well you get along with other people. Competence incorporates knowing the business and being able to apply that knowledge in the form of good business judgment. There's also a relative side to competence, since your work will inevitably be compared with others'.

Your goal with regard to competence has two basics:

- Become a key contributor by focusing on results that are important to your organization.
- Develop yourself professionally.

TECHNICAL SKILLS AND HARD WORK

Strong technical skills and hard work are the foundations on which you will build your career. It's difficult to separate the two, for it's the *quality and quantity of your work* that the management will be most interested in. In other words, *accomplishments and results that create value for your organization are important.* As we discussed in Chapter 1 and other earlier chapters, setting objectives and aligning them with your management's objectives will focus your skills and diligence into the right channels.

How you get those results is also important. That includes how effective you are when completing your assignments and selling your ideas. In addition, results can't be obtained at any cost—they must be achieved with integrity.

We'll develop those two ideas—effectiveness and integrity—later in this chapter and the next. But first, let's consider developing your technical skills.

Technical Skills Development

Your technical skills were first developed in college. You probably studied engineering (although not necessarily), which gave you a good frame of reference for the popular engineering disciplines (such as chemical engineering, civil engineering, electrical engineering, mechanical engineering, and more) and gave you a concentration of skills in your major.

As you move into the business world, your assignments often involve skills you haven't learned yet. Don't let that disturb you—it's normal. Every industry and every company have specific technical areas that are important to them. This will cause you to crack the books or get some help from an expert to learn what you need to know. Sometimes there are company courses or manuals available to help. This learning process is analogous to mastering the use of a new tool. You go to the hardware store and buy a sander to refinish a piece of furniture. You will learn how to use the sander, and the next time you need it, it will be available, in your tool kit, to help you do your work more efficiently. As a project engineer, you are also able to draw on other project people to help do a job or solve a problem, and you will learn from them in the process.

The development of technical skills is an ongoing process for project engineers, since such a large component of their work is, indeed, technical. As we discussed earlier, you learn the new skills you need today, but you seek training to prepare yourself for future assignments or to give yourself more perspective on the jobs of other persons that work alongside you. You may also seek continuing education to keep abreast of new developments in your field. This could involve reading technical or trade journals. Many companies have technical training courses to develop specific job-related or business-related competence. Some companies offer educational assistance programs to encourage continuing education, like night school.

Career Development

After a few years, when you have filled most of the gaps between the technical skills you have acquired and the skills required for your job, you're considered to be technically competent. Once you've become a competent engineer, your career can take off in several directions. One natural avenue is to stay in your current job and build the deeper experience and judgment needed to *become an expert*. Another way is to advance to the level of *project engineer*, where you can have more opportunity to influence the business by directing the efforts of others. This could lead eventually to a supervisory position. Yet another career path is to *transfer to a new job*, which provides other technical or business skills to round out your experience in the industry you have chosen. For example, you could move into a cost engineering position, a construction assignment, or a corporate planning job to learn something completely different. The possibilities will depend on what positions the organization has available at the time you are ready for a new assignment and your own skills, interests, and aspirations.

Opportunities materialize unexpectedly, which is why having a plan for your career and discussing it with your supervisor are so important. *It doesn't have to be an elaborate plan. A couple of job positions you would like as your next assignment and a general idea of the path to your dream job is sufficient at this stage.* During your annual performance discussion with your supervisor, you can ask what the organization has in mind for your next job. Even in smaller companies, where the process is less formal, your career is mainly steered by you and your immediate boss. Conversations with your boss should naturally lead to an opportunity to express your interests and career preferences. Letting your supervisor know of personal considerations, family constraints, or a strong preference to live in a certain location (or avoid other locations) helps ensure those factors will be taken into account.

Unexpected opportunities can also come in the form of job offers from other companies or from layoffs that force you back into the job market. Knowing what you want to do is a key ingredient in your job search decisions.

PERSONAL EFFICIENCY AND EFFECTIVENESS

In parallel with developing your technical competence, you can enhance your performance by concentrating on your personal efficiency and effectiveness. As mentioned earlier, hard work contributes to your quantity of work. Efficiency gives an extra boost to the quantity of results you can achieve. Effectiveness, on the other hand, relates to the quality of your work and its value to your organization.

What can you do—as a new employee in the workforce—to make yourself more efficient and effective? You'll be glad to know there are many things you can do. Here's where to start.

Getting Organized

Getting yourself organized is probably the first order of business for a new employee. It's a basic way of approaching the job that involves

- keeping track of the tasks you have to do,
- getting to the right place at the right time,
- processing the flow of information that comes your way,
- keeping track of the results from the meetings you attend, and
- reminding yourself of important deadlines or other items.

A lot of the techniques for doing those functions are built into your organization's email system. Many people use an electronic personal organizer that can communicate with their PC. This technology will no doubt develop further, and you'll want to keep up. Others prefer a more manual system to keep track of their tasks and calendar.

All of this, however, is simply technique. The real issues are managing your time and concentrating on your highest-priority objectives and tasks. A short course on time management will give you many valuable ideas, but here are a few tips to hold you for now.

Task List

Consolidate the tasks you have to do onto a single task list. You only have one life, so you only need one list for business and personal tasks. Have an approach for designating high-, medium-, and low-priority items.

Calendar

Keep your calendar up-to-date. Hopefully, it's tied to your email system so that you'll automatically receive meeting notices. Include both business and personal commitments if your organization allows that. Consider your priorities when committing to attend a given meeting—do you really need to be there?

Processing Mail

Particularly on projects, you will be handling a considerable amount of mail—both email and paper mail. You'll probably find that you will get most of the value from a piece of correspondence when you first read it, so I suggest you make the most of that initial reading. The most efficient people try to handle a piece of mail only once—answer it on the spot, delegate an action or a response to someone on your team, circulate it for information, file it, create a reminder, make a calendar entry—and get it off your desk. If you can master that talent in your first year, you'll boost your efficiency for the rest of your career.

Again, screening your mail and establishing high, medium, and low priorities is a way of ensuring that you're working on the most important items first. We all know the saying, "Anything worth doing is worth doing well." However, I have a friend who coined the corollary, "Anything not worth doing is not worth doing well." Choose where you spend your time wisely and don't work on trivia. When assigning priorities, project engineers should consider the tasks that they delegate to others as high priority. In that way, your team is working on the delegated tasks while you are working on your tasks.

Early in your career or when you change jobs you may find it difficult to distinguish what is important and what is not. Also, sometimes you don't have a clue what to do with a piece of mail. When that occurs, it's handy to keep a deep drawer in your desk available to stack all the low-priority paper mail that you just don't know how to handle. If someone comes around later looking for a response to one of those items, you can confidently retrieve it, clarify the requirement, and promise to supply what they need.

Email

Email is a fantastic asset for communicating in our global business community, but it must be used with judgment and caution. First of all, email is forever. Any note you write can be retrieved, so be careful what you say. Also, I've seen carelessly written email sail around the world in a matter of hours and find precisely the person that the writer wished it hadn't. Along that same theme, don't email over your boss's head (or bosses' heads).

One final cautionary note about email: don't use it to address situations or issues involving conflict. In those situations, a face-to-face meeting or a phone call is always preferable to an email, which can be misunderstood. Resolve the conflict first, and then, if necessary, respond by email confirming that resolution. The same applies to other written correspondence. Solve the problem through discussion and then document the solution with a letter or memo.

Meeting Journal

Many people in all walks of life find it useful to keep a journal in order to capture the important points from the meetings they attend. It serves as a handy chronological reference for information and the tasks you have agreed to undertake. You can transfer the tasks to your task list, if you like, or periodically review the journal and check off your action items when they're completed.

Call-up File

A call-up file is handy to jog your memory. It can be as simple as a hanging file (or an electronic equivalent) with 31 dividers in it—one for each day of the month. You can put any kind of paper in it under the day you would like to be reminded. The catch is to remember to check it every morning and to look ahead if you are going to be away from your office.

Being organized is necessary if you are going to be effective, but organization can be overdone. You will have to find your own optimum on the scale between doing nothing and becoming a compulsive desk-straightener. If you get yourself organized within the first few weeks, you'll soon develop the reputation for being dependable.

Initiative

Like being organized, initiative is another basic way of approaching the job. It's a mind-set that aligns your objectives and actions with those of your management. You've probably heard expressions like "She's a self-starter" or "He works independently." People with high initiative are seen as *handling their job without needing much supervision.* They understand what's required of them but that doesn't mean they're working in isolation. They know how their work fits with other activities, and they communicate effectively, producing results that are useful to the organization.

In addition, they look at the work ahead and *anticipate* what they should do to accomplish their tasks and goals. The better performers also look at their responsibilities and anticipate what their bosses need in the way of information or results and provide it in a tactful and timely manner.

Edgar's answer (in Chapter 5) to Ron's question about *expanding your role* gives additional insight into initiative:

> It's easiest for you to just handle your own job. But what I'm asking from each of you is to take on more responsibility. If there's a problem, take a leadership role in finding the solution. If there's an issue that cuts across other areas' or even other companies' work, pursue that issue until everyone involved agrees on a solution. If some task needs coordination, step up and volunteer to do it.

Initiative is one of those intangible aspects of effectiveness. It's questionable whether it's a skill or an innate attribute, but one thing is certain: it differentiates the top performers from the rest. If you find that you have initiative, develop it as you seek to improve your effectiveness.

It boils down to these simple steps:

- Handle your job.
- Anticipate what you and your boss will need and do it.
- Expand your role.

Communication

Communication creates effectiveness. Both the organization and you need effective communication to succeed.

Chapter 5 illustrated Wizard Oilfield's organizational interests. Sara's four points (communicate early, often, informally, and candidly) are essential in a high-performance organization. The people in the organization need that exchange of information to function at the highest levels of achievement. Sara's advice can help you and the rest of the organization communicate effectively in your daily work. If the information doesn't flow to the people who need it, the work will be inefficient at best and disastrous at worst. In our Rio Bend compressor case study, only money was wasted by poor communication. But if the lack of information had resulted in a faulty design, the consequences would have been much more severe.

Your ability to effectively communicate will also serve your personal interests. Communication is the way you will influence the organization and, to a large extent, the way the organization will perceive you and your work. It can be a major contributor to your personal effectiveness.

Conversations

First let's focus on important conversations. Few people recognize the benefit of planning their everyday communications with others. Of course, they will organize their thoughts and painstakingly agonize over every word in an important letter or take weeks to prepare for a presentation. But few will spend the time to think through even the key points for a vital conversation with their boss or team member. For a good example of this technique, let's flash back to the first scene in Chapter 5 ("September 20: Sara Sends Up an SOS"). When we join the story, Sara has already organized her thoughts and has landed on a solution. She then gets her reasoning straight and writes the main points on a piece of paper. Her business judgment tells her that she needs to have a recommended candidate and an organization chart ready. Instead of writing a memo or choosing a face-to-face meeting, she elects to handle her conversation with her boss over the telephone— probably because she feels confident in her relationship with Kramer. The setting is late in the day, when people are generally more receptive and less distracted. Because of a schedule problem that Kramer was managing on another project, Sara expected that Kramer would be amenable to avoiding the same thing on her project. She also delivered the message in an interpersonally sound and confident manner. Keep this approach in mind the next time you have an important conversation. *Get your thoughts organized, plan the message, choose the setting and timing, and then deliver.*

Writing

Writing is a fundamental skill for effective communications. In business you will routinely write memos, letters, reports, proposals, work plans, and a host of technical documentation. This is the way you will sell many of your ideas.

Effective writing in the business world borrows from journalism. A time-honored structure for a letter, memo, email, or even a report will often have four parts:

1. *What's the news?* Often your written work will make a recommendation or draw an important conclusion or convey a business proposal. State it clearly and succinctly at the beginning of the document. If you can, tie your message to an important business objective. You may find it necessary to give some background information to get your reader "on the same page with you" but don't let that get ahead of the news or drown it out. Some people find this approach difficult. They want to build a solid case before exposing the news on the last page. However, building the case is the purpose of parts 2 and 3 of the piece of writing. Your management will be looking for the news up front.

2. *Why?* Here is where you sell your idea in a mature, reasonable manner. Give the most compelling reasons and incentives first. Keep it short and logical. By this point you should have given your reader enough material to form an opinion on your proposal.

3. *What is the supporting information—your reasoning and the discussion of facts, figures, issues, risks, alternatives, and timing considerations?* Now you can make your case. Lay out the information and documentation that support your point. You may want to provide a more complete or a more detailed summary of your proposal. If you have considered alternatives, summarize their advantages, disadvantages, and the reasoning that led you to your proposed course of action. Important issues should be addressed and risks should be pointed out, along with possible mitigation. Also, any important timing considerations, such as deadlines or times when an opportunity will disappear, should be presented. And, of course, present the relevant cost impact of your message. In other words, lay out a balanced, credible argument that will sell your idea and keep the amount of detail appropriate for the audience. The lead project engineer will need a different discussion than senior management. Include only the information necessary to convince your audience. An extraneous point could unsell your case.

4. *What are the next steps?* An effective way to end the document is to propose the next steps. If you are recommending something, the next steps would include the activities and schedule for implementing your proposal.

Presentations

Like writing, presentation skills are an essential key to effective communication. In fact, presentations may be more important because of the potential exposure they can give you to management and key individuals in the organization.

In business presentations, the message must be crystal clear, and you should always have a recommendation. It's wise to lead with one chart that captures four essential elements that the audience (usually management) wants to know:

1. What is the purpose of this presentation? What do you want me to agree to? What are you recommending?
2. Why should I do this? What are the incentives?
3. What are the problems or issues that might arise if we do what is proposed (or don't do it)? Can they be mitigated?
4. What are the next steps and their timing? When do we have to act to capture this opportunity?

This way your whole story is on one page and it's the first thing you cover. If management arrives late and you have to shorten your presentation, you can cover the first chart and then hit the high points, while skipping the less important material. If the meeting goes in an unexpected direction or gets hung up on a single point, you can return to that chart to try to get things back on track. If the manager or vice president is called out of the meeting to handle something urgent, you can use that chart and conclude with your recommendation before he or she leaves.

Here are a few hints to keep in mind while preparing for presentations.

- Know your subject.
- Know your audience.
- Figure out the message and organize your thoughts around it. Keep the talk interesting to the audience.
- Prepare slides, charts, training aids, and speaking notes, as needed, to help get the message across.
- Review the material with your boss and take her or his comments into account.
- Rehearse the talk out loud, until it's yours and you're comfortable with hearing your own voice. If possible, rehearse it once with the equipment and in the place where you will give it. If that's not possible, at least familiarize yourself with the place and equipment. Think about possible equipment malfunctions and how to recover from them.
- Anticipate questions and think through your answers to them. Make sure you understand a question before you try to answer it.
- Present the talk with confidence and genuine conviction.
- Stay within your time commitment.

Listening

Some are surprised to learn that communication involves receiving information, as well as giving it. We all know that communication is a two-way process and yet many people ignore listening, which is one of the key aspects of effective communication. Let's start with the basics—the well-known communications model.

When people speak, their ideas are framed in words that are derived from their background and experience. As they speak, their body language, expressions, and emotions all convey meaning. When you hear their words and take in to account how they deliver them, you are essentially interpreting what you hear and see, based on your own background and experience—which are different from those of the speakers. At best, the task of conveying an idea—intact—from one brain to another is challenging.

As you add more participants to the conversation, the challenge increases exponentially. If someone adds duplicity or lying (as Chet did in Chapter 5) communication becomes even harder. If the participants in a conversation have cultural differences, mutual understanding takes on a whole new dimension of difficulty. Those are all excellent reasons why you should listen well—to get the message. As you'll find later, it's also the first step in understanding the politics in your organization.

Your purpose, as a listener, should be twofold:

- To hear what is being said.
- To try to understand the thought that is being conveyed.

Both of these points are difficult, but the first is the easier of the two. It just takes careful listening. The second requires subjective judgment.

In a conversation, probably the best tactics for understanding a message are to

- ask clarifying questions, and
- give feedback to state your understanding of what was said and hopefully confirm it.

In other words, actively participate in the conversation, even though you're primarily listening. After (or even before) the conversation, you can try to verify the information by checking with other sources. It will give you some insight into whether you are receiving incorrect statements.

If a conversation is charged with emotion, try to defuse it before continuing to exchange ideas. Recall Edgar's encounter with Walter in Chapter 5 ("October 1: Edgar Jump-Starts the Engineering"). Walter was upset with Edgar's more rigorous approach to planning. He also may have felt that Edgar was talking down to him. This was a rather mild disagreement, but Edgar took it seriously. He called a break in the meeting and probably had a one-on-one conversation with Walter to explain the rationale for the new approach. Whatever he said satisfied Walter and brought him back into alignment. There are books and courses on conflict management, but a lot of it is simply sound interpersonal skills, patience, common sense, and respect.

Attitude plays a central role when communicating, whether speaking, conversing, or listening. If you start with an attitude of respect for the other person, the respect will come across in your body language and the words you speak. Most people will respond positively to this. Some won't.

Selling Ideas

Regardless of the manner in which you communicate, the strength of your argument should lead the audience irrevocably to your conclusions. A weakly thought-out position can't be sold. It should contain the analyses, facts, and logic that lead to your conclusions, and it should be delivered credibly and convincingly. Counterarguments and issues with the proposal should be dealt with, but extraneous information that can cloud or confuse your conclusions should be removed, even if you find it interesting. If you're new in the workplace, it's easy to dwell on topics you're interested in rather than what the organization is interested in. *Staying focused on the business objectives will steer your communication in the right direction.*

Effective Meetings

Conducting effective meetings is another project engineering skill that's closely tied to leadership. It's a way of making things happen. You will want to call meetings to solve problems, kick off activities, orient your team, hold team-building sessions, inform clients, or brief management to inform them and get their guidance or decision.

Many people feel that meetings are an incredible waste of time and, indeed, some are. However, effective meetings create results in three ways:

1. People accomplish work getting ready for a meeting.
2. The meeting itself focuses attention on a specific purpose or result.
3. Responsibilities for additional work are often delegated, agreed on, or otherwise established during a meeting.

Organizing and conducting effective meetings also will reinforce your stature as a leader.

The starting point for an effective meeting is a *purpose* for the meeting, which you will expand into a *set of objectives* and an *agenda*. You can use Edgar's planning meeting as an example of an effective meeting. Recall that Edgar had both an objective (Edgar may have had other objectives, such as team building and delegating tasks, that he didn't tell the team) and an agenda. Those are the minimum requirements for an effective meeting.

Based on the objectives and agenda you can determine who should attend the meeting. There are two general classes of people who should be there:

- Those who can contribute essential skills to the meeting (especially in a problem-solving or brainstorming session)
- Those who need to be committed to the results of the meeting so that they can implement them

Any other attendees are probably redundant unless, of course, there are political or other cogent reasons for them to be there.

Once you have nailed down *what* and *who*, you can estimate the duration of the meeting and set the date, time, and place. Meetings with peers or other project members can be called on relatively short notice, but if you want them to be prepared for the meeting, give them time and some direction about the preparation that's needed. If several members of senior management are involved, you may need weeks of lead time to get a quorum.

When preparing for the meeting you can organize any material you will present and possibly prepare charts or handout material that contains the message you want the participants to take away. Remain focused on the purpose and objectives. If you're presenting a proposal to management, it's best to have a recommendation. Your organization will have its way of presenting material at meetings (handouts, slide presentations, overhead projection, and more) and you need to get in step with that. If others will be presenting information and you are in charge of the meeting, do what you can to ensure they prepare themselves.

During the meeting, you will facilitate the discussion and try to keep it focused on the objectives and the agenda. Again, use Edgar's meeting as a guide. From time to time conflict is bound to arise. Treat people with respect. Listen to their viewpoint and try to understand it. If possible, take their viewpoint into consideration as part of the group's solution, even if it just means acknowledging it as a credible point of view.

At the conclusion of the meeting, you should facilitate a wrap-up session. People in meetings hear different things or see things differently. Some daydream or doze off. If an important conclusion is reached, write it on the whiteboard so everyone can see it and agree or disagree on the spot, while all are assembled. The minimum wrap-up will be going over a list of action items that you have been writing on the board or flip chart during the course of the meeting. The maximum would be writing minutes, which are distributed as an email, memo, or letter. Some companies require written minutes, especially if commercial commitments are being made.

Of course, you can do all of this and still not achieve your meeting objectives—through no fault of your own. Some supervisors or managers, usually at the lower levels, avoid making decisions or postpone them until it is safe. The project engineer can lay out a crisp recommendation with airtight supporting facts and reasoning but gets no response from the boss. This certainly puts the project engineer in a high-risk situation of being forced to act based on his or her own judgment. If the decision needs the boss's financial authority, you have no choice but to keep following up with the boss until you get the necessary approval. If it doesn't require financial approval, the situation may require that you continue the work as you recommended. In either case, it's desirable to document the situation with minutes of the meeting so that you have something to rely on later if things go south.

Finishing the Job

For some very competent engineers, finishing a job is a problem. This can be due to a lack of clear objectives, which can be dealt with by making the objectives more specific—with completion dates and measurable results that show when the objectives must be achieved.

Visualizing the end product is another tool to help the engineer prepare a meaningful end product. For example, outlining a final report before the project begins may help guide the work to completion. The outline will undoubtedly change as the work develops and progresses but it provides direction and focus on the end point.

A frustrating situation for both the engineer and the company occurs when an extremely hard-working engineer, with excellent technical skills, can't distinguish when the job is technically finished. Either out of interest or conscientiousness, the engineer continues to chase technical issues instead of making judgments and completing the work. If this happens time and again, the engineer's effectiveness will diminish because he or she never finishes anything. One way of addressing this situation, is for the engineer to seek the advice of a respected colleague or mentor. The additional viewpoint may help remold the engineer's technical judgment and improve his or her business judgment over time.

Teamwork and Leadership

In Chapter 3, we considered several aspects of team building, teamwork, leadership, and management. Here we'll put teamwork and leadership into the context of a project engineer's personal effectiveness.

At the root of effective teamwork and leadership is the ability to get along with others. Your effectiveness as a team member, team leader, or project engineer will depend on the rapport you have with people and the respect they have for you, your opinions, and your leadership.

Teamwork

In some organizations, effective teamwork is a primary business objective that ranks with safety, quality, cost, and schedule. As an employee, it's essential that you be able to work effectively as a member of a team. As a project engineer, it's crucial that you be able to lead a team.

Teams can be an organizational unit, a task force, or group convened for a specific purpose—for example, to solve a problem or investigate an incident. As a project engineer you will be involved in all types of teams. Some essential success factors for effective teams are the following:

- Selection of a leader
- Alignment of the team members with the team's objectives
- Agreement on how to make team decisions
- Self-assessment of the team's performance

Let's consider those in more depth.

Generally the first order of business for any team is to select a leader, if one has not already been appointed. In unstructured settings, the leader may emerge from the team members to fill a void in leadership.

Second, the team should go through some process to align the members with the objectives for the task at hand. Even if the team is given a formal charter or the objectives are imposed by the leader or the organization, it's still important for the individuals of the team to discuss, clarify, and, if possible, add value to the objectives. If the team doesn't accept the objectives as somewhat consistent with their individual goals, they may perform in the short term but become dissatisfied in the long term and perform less effectively.

Effective team performance is enhanced if the team has an agreed way of reaching decisions. Group consensus is generally the aim. To reach consensus, members should feel free to express their opinions. The team objectively discusses and debates the various opinions and issues as they reach unanimous agreement on the best solution or course of action. The leader facilitates the discussion and keeps the team on track. The team reaches a consensus when everyone accepts a given solution as being the best. It may not be each person's preferred solution, but it is the best solution when all factors are given their proper weight. For example, the solution to the gas compressor's inlet-outlet problem in Chapter 5 was not the best solution from the structural discipline's point of view, and it was certainly a problem for the control system project engineer. Also, the client had to pay more money. But overall, it was the best course of action for the project.

Finally, the team needs to take stock, occasionally, on how it is performing as a group. Are they getting the job done well? Do the team members feel included in the team and satisfied with how things are going, or are there contentious issues within the team that need to be resolved? Is the leader dominating the conclusions?

Being able to work effectively on teams is such a fundamental part of the project engineer's job that it should be practiced and perfected. There are good seminars on team development available that teach the fundamentals of interpersonal interactions and effective team processes. The seminar should include group exercises where the participants have the opportunity to function both as group members and as the group leader. One of the most enlightening parts of this type of exercise will be the observations and feedback that the participants receive on their performance from their classmates. Attending a teamwork seminar can be an important first step toward becoming a leader.

Leadership

Effective leadership, from the project engineer's point of view, is balancing three main factors:

- Getting the job done well
- Taking care of the people
- Managing the situation

We've considered the first point in the earlier chapters and you're probably beginning to feeling comfortable with that role by now, but the people side may deserve more explanation in its application to leadership. If you accept the definition that management is influencing people to accomplish organizational goals, then your effectiveness hinges on how well you influence those who look to you for leadership.

Project engineers are more likely to be effective leaders if the people on their team respect their leadership, are reasonably well paid, find the job rewarding, and feel their individual career objectives are being furthered by doing a good job. The art of motivation is a big part of that. Leaders must get to know each team member and what motivates them. It's best to recognize one fact from the start—not everyone is motivated by the same things that the leader is.

The first step in managing people is to establish rapport with them. But there is an ongoing need to keep the lines of communication open and resolve issues as they come up—whether they are individual problems or conflicts between individuals. Listening is one of your primary tactics to understand what's going well, what's wrong, and which issues should be addressed.

As discussed earlier, goal setting can help align the team. If the team participates in establishing the goals, they will be more motivated to do the task and feel like their own goals are being advanced in the process. Project engineers can interject their own perspectives since the leader's high expectations usually lead to high performance by the team, just as low expectations lead to low performance.

Part of taking care of the people is enriching their job. For most engineers, this occurs naturally as they are given more responsibility, increased scope to their work, and more challenging assignments.

The third factor in leadership—managing the situation—adds an interesting twist, since each situation is usually different in one way or another. Because of this, most supervisors and managers would agree that there is no single style of management that applies to all people and in all situations. What works for a new employee may not work for an experienced expert. The techniques for solving a technical problem will be different from those of dealing with a crisis, like putting out a fire (either literally or figuratively) in the office. It all comes back to the art of motivating people to perform tasks in differing situations. In addition, there are situations where something more than good interpersonal skills is necessary in order to lead effectively.

Power and Influence

Project engineers have several types of power available when interpersonal skills aren't enough to carry the day. Consider a situation where the majority of your team has an opinion different from yours about how to fix a technical problem. If you have information available to you that they don't have, you can use that information to give a persuasive argument that convinces them your solution is correct. Or if you are an expert on some crucial part of the problem, you could add that evidence to prove your point.

Sometimes you'll encounter individuals who are being adversarial or devious in order to accomplish their own goals. You try to use your best interpersonal skills, but the person maintains a stance and refuses to budge. It will take power to move that person from their stance.

Hersey and Blanchard (1982) define power as "the potential for influence." This implies that power is a resource that a person may, or may not, choose to use. They point out seven sources or bases of power:

- *Coercive power*, based on fear, intimidation, threats, or punishment
- *Position (legitimate) power*, based on assignment to an organizational position of authority
- *Expert power*, based on expertise, skill, and knowledge
- *Reward power*, based on the ability to grant pay, promotion, recognition, or other incentives
- *Personality (referent) power*, based on personal traits and attributes that others admire and respect
- *Information power*, based on possession of or access to valuable information
- *Connection power*, based on having connections to influential or important people

Three of those power bases stand out as being available to a project engineers: expert power, personality power, and information power. New project engineers probably have little access to coercive power, position power, or reward power, since they generally won't have direct supervisory responsibility. On the other hand, Edgar, the Lead Project Engineer in our story in Chapter 5, had these types of power available because of his supervisory position and by virtue of the confidence his management has shown in him.

We actually talked about expert power earlier when we referred to the project engineer's technical expertise and specific expertise regarding her or his area. Because the project engineer is an expert on those subjects, others will be influenced by what he or she says and does. The inexperienced members of a project engineer's team will likely take direction and suggestions without much resistance or reluctance. However, more experienced team members may not be as responsive to the project engineer's expert opinions. They may be irritated by a directing management style and may respond better to a style that seeks advice and delegates tasks, leaving the details for them to work. Expert power can also promote the project engineer's influence with both management and peers, if used wisely. They will formally or informally seek expert advice, and in the process the project engineer will have influence.

Personality power probably needs no additional explanation. It's available if the project engineer has the traits and skills to get along with others. If this is a rough area, the project engineer can seek training to improve his or her interpersonal skills or, at least, keep them from becoming a negative influence.

Information power is similar to expert power. It gives the project engineer influence with subordinates, peers, and management. Project engineers with good formal and informal information networks are on top of what's happening or about to happen. If others think the project engineer has inside information, they'll come seeking it, which will give the project engineer influence. The project engineer is the natural conduit for information flowing to his or her team. Keeping team members informed of necessary information on a regular basis—especially through meetings—will strengthen the project engineer's information power with them, and his or her stature as a leader. Regular communication with bosses and peers will usually have the same effect. Remember: communicate early, often, informally, and candidly with team members and bosses.

Connection power can also be available to the project engineer as she or he gains exposure and develops contacts with powerful people in the organization. We'll talk more about this type of power in the section on office politics.

As you learn to use power, you'll find that you can influence meetings, affect important management decisions, and even twist arms, while still being interpersonally competent. As you grow in responsibility over your career, you'll find many other ways to wield power. But power can be allusive. It can slip out of your grasp if you don't use it wisely. The days of absolute power being vested in a certain position are gone, if they ever existed at all. Even position power will wane if the leader filling the position doesn't maintain an effective relationship with the management and the people being led.

After all, power is just an extension of the idea of influencing others, and the most effective sources of power come from your relationships with those above and below you on the organizational ladder. As you grow in effectiveness, you'll find yourself using the various power bases continually to get your job done. And when you feel like you are being pushed around, you'll look for an available source of power to accomplish your objective.

Training

You may have sensed that we've barely scratched the surface on the topic of personal effectiveness, but those skills are enough to get you started. As I've mentioned a few times, training is crucial. It's a shared responsibility between you and your supervisor. In the short term, getting the right training, early in your career, will make you more productive and help you avoid mistakes. In the long term, you will benefit from and build on the techniques you learn for the rest of your career.

Without being a pest, you should try to discuss training with your supervisor. Perhaps a good time to start is during your job interview and when you report to work. As you engage in this dialog, keep the list of seminars and workshops in Figure 6.1 in mind. The order of the list is the recommended sequence. However, if you have trouble with public speaking, move it up to the top of your list.

The best way to enhance your speaking ability in a short amount of time is to attend a course or workshop where you will be able to practice your public speaking skills as you learn them. Take an in-house course if it's available, since it will be tailored to your specific business environment. Some people have found

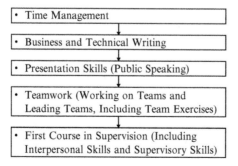

Figure 6.1   Recommended seminars and workshops for new engineers and project engineers.

Toastmasters International helpful in developing their speaking skills. Check their website to see if it's right for you (www.toastmasters.org).

The American Management Association (AMA) presents seminars on all these topics in major United States cities (their website is www.amanet.org). Your company or organization will generally not send you to a supervisory course unless you have demonstrated some potential for becoming a leader. If you don't get that opportunity, you may want to request an AMA seminar on interpersonal skills to enhance your effectiveness in dealing with others.

Keeping an Awareness of Your Effectiveness

The final thought on effectiveness is the idea that you should maintain an awareness of your effectiveness as you engage the business world. This means continually working on becoming more effective in how you do your job. It means listening to the wise counsel of your supervisor and doing something about it. It means reflecting on your own performance. Sometimes after a particularly good or bad experience, it's useful to reflect on your effectiveness in the situation and, if necessary, how it could have been improved. And you can keep those skills we've discussed in mind and develop them.

BUSINESS JUDGMENT

Building business judgment is an elusive part of becoming competent and developing professionally. In addition to knowing the technical side, you must strive to learn about the business. Even as a new employee, it's necessary to find out what's going on around you and especially how the management thinks. Sources of information include group meetings or orientations, publications, lunchtime conversations, and company or industry websites. One of the most important sources can be conversations with your supervisor. All of these will give you information on the company or organization and its business.

But as you learn the business your goal is not to accumulate a wealth of facts. You're trying to figure out the structure of the company, its business objectives, its main players, how the management thinks, and how the company works. Start with understanding your own organization and then branch out to other units. You're trying to build conceptual skills that will help you put your job into the overall company perspective.

Does this make sense? Consider a situation where you are giving a presentation to your boss's boss or even higher. For instance, as in Chapter 5, you could be Ron Neuman presenting the change request package to Sara and her boss, Kramer. You'll need to give them accurate, credible advice that will meet their expectations.

Your technical competence will help you keep your facts straight, but it is your business judgment that will make your presentation credible and relevant to their needs. In this case, both Sara and her boss will want to know if the estimated cost for the change is believable. Knowing this, you can avoid dwelling on a needless, detailed explanation of how you prepared the estimate, and instead focus your presentation on a comparison of your estimated cost with benchmark data from other successful jobs. That will give Sara and Kramer confidence that the costs for this change order are in line with what is normal. Also, they will both be worried about avoiding the schedule penalty. You'll have to address that by estimating the likely schedule slippage and giving them a recommendation (which you have reviewed with Edgar and then Sara ahead of time) on how to control that risk.

The compressor case study brought out the importance of commercial and contractual issues on projects. This is an important part of business judgment to which some young engineers are often blind. As a project engineer, you will be wise to develop a relationship with the contracting and purchasing personnel at your level. Cooperating with them will help both of you succeed and will sharpen your commercial understanding and skills. Remember that managing the risks and managing the money for your area will require technical, commercial, and business skills.

As you advance, business skills are nearly as important as technical skills in shaping your overall competence. A senior executive that I know advised people to objectively analyze

- what worked,
- what didn't work, and
- why.

As you repeat that process, your business judgment will develop.

But business judgment doesn't come overnight. It will emerge gradually, as you build your business knowledge, take a few lumps, gain experience, and absorb sound coaching and advice from your supervisor. You'll begin to see ways to have more influence on the success or profitability of your organization. You'll see business opportunities and have the understanding of how to sell your ideas to management, in a convincing way. Your own results will become more aligned with the objectives of your organization, and your contributions to those objectives will be more significant.

PERFORMANCE EVALUATIONS AND THE COMPETITION

Every organization, large and small, will have some way of evaluating their employees to deciding how to allocate salary increases and promotions. In small organizations, the evaluations may be subjective and informal. In large organizations, performance appraisals will usually be somewhat more objective and systematic. Whether the evaluations are formal or informal, the organization tries to assess an employee based on four factors:

- Results and accomplishments
- Value of the employee's contributions to the organization
- How well the employee does the work
- The degree of difficulty of the work

Let's review these more carefully.

You will recall from the discussion of technical skills and hard work that relevant *results and accomplishments* are important to management. Technical competence and hard work can distinguish your performance from that of your peers, especially early in your career. As you develop and acquire business judgment, your achievements will grow in importance.

The *value of an employee's contribution* to the organization is closely related to results and accomplishments. Those results that enhance the profitability of the company or make significant contributions to organizational goals are valued more highly.

Evaluation of *how well an employee does the work* essentially measures and weighs the aspects of personal efficiency and effectiveness that we have covered above. High on the list of effectiveness skills is usually communication and the ability to get along with people. Effectiveness is something the supervisor will see on a day-to-day basis. If an employee performs well in an important meeting or gives a skillful presentation, others will notice. This is why communication skills are so vital to success. The bosses will know effectiveness, results and accomplishments, but everyone will see effectiveness and the word will spread.

The *degree of difficulty* of an employee's work is related to the nature of the job rather than the employee's contribution, skills, or attributes. It's somewhat analogous to a diving or gymnastics competition. Someone who does a difficult job well receives more credit than someone who does an easy job well. Another way of looking at it is that those who hold more responsible positions may receive a higher ranking in the evaluation process. This might seem unfair, but it certainly underscores the need for the employee to be involved with his or her supervisor in career planning and to express preferences for future assignments.

So what should you do about competition and evaluations at this stage in your career? It's best not to be overly concerned with competing. Be aware of it, but don't dwell on it. Instead, spend your energy where it can make a difference. Work on your competence by concentrating on the more tangible things like becoming technically competent at your job, working hard, focusing on priorities, building effectiveness, developing business judgment, making friends among your peers, and getting along with others. If your supervisor coaches you about something, listen to him or her and fix the problem. Establish yourself as a key member of the organization who contributes to business objectives and you will compete.

OFFICE POLITICS

When I was in Canada on a business trip, the title of a newspaper article caught my attention:

When Good Isn't Good Enough
Becoming a master of office politics doesn't mean you have to compromise your principles and ethics. But you ignore the minefield at work at your risk (Young, 2003).

That certainly sets an ominous tone for the subject of office politics, possibly more ominous than it should be for a recently hired engineer. Let's take the author's warning by watching where we step, but proceed on a more positive note. We will, however, come back to the negative side of politics later, because it can't be avoided.

Let's start with a formula for success that some people have found accelerates their career. You begin with competence as your base. All the aspects of competence we have just discussed will help you do your job well. You develop your technical skills and work hard. You increase your effectiveness by getting the right training, becoming a team player, developing your leadership skills, and learning to wisely use power and influence. You build your business judgment.

So what's missing? The answer is *patrons* and *exposure*. Your success comes not only from your competence but from the perception that influential people have of your competence and your potential for advancement. That's where the politics come in.

It's not the engineer who always sits in his or her office and turns out the work that gets ahead the fastest. It's the one who associates with and is exposed to higher management in a positive way—in addition to turning out the work.

It goes like this. You keep your management sponsor or patron informed by giving correct, credible information and advice. You keep your confidence and credibility high and give your patrons the results they need, when they need them. They succeed and you succeed. And when they need someone for an important task or position, you will be considered for it. The result can be dramatic jumps in your career.

Of course, this approach doesn't work for everyone, but, at least, everyone should be aware of it. Now that we have the basic formula in mind, let's go back and fill in the details.

PATRONS

Patrons are your management sponsors. The most obvious potential patrons are your supervisor and members of your line management. However, other members of the management team in your organization, influential technical leaders, and former bosses can also be patrons. As you make contacts and connections with those types of persons, add them to your information network.

Let's return to Chapter 5 for an example. After Ron Neuman succeeded in solving the nozzle problem, both Edgar and Sara had most likely moved into the camp of Ron's patrons. These two are probably the most natural patrons Ron could have. They are his direct supervisor and manager. As he continues to do a good job, his connection to Edgar and Sara will likely grow. Ron will have most of his contact with Edgar, but occasionally there will be briefings or conversations where Ron will have the opportunity to directly influence Sara's opinion of him. However, he can't be seen as "going around" Edgar to curry Sara's favor.

Ron has possibly developed other sponsors in the short time he has been with Wizard Oilfield. Walter, Ron's first mentor, could be one, and Bill, the Engineering Manager who has stewardship over Ron's career, could be another. Even peers can become patrons later on, so don't forget them. Ron will likely include all of these potential patrons in his network by contacting them from time to time when an opportunity presents itself.

Ron can also branch out to other parts of the organization in his efforts to cultivate patrons. For instance, he might informally go to the Purchasing Supervisor's office and ask him or her who would be the best contacts in the Purchasing Group concerning his compressor package. If handled well, this type of interaction can leave a positive impression of cooperation and teamwork, while giving Ron valuable information and contacts.

EXPOSURE

Exposure can come at any time and in a number of ways. It can be either good exposure or bad. Once you become aware of this, you can be prepared to make the most of those opportunities or mitigate any potential damage. Sometimes fading into the background or leaving the room will avoid negative exposure.

The backdrop is how you project yourself within the organization, on an everyday basis. In their book, *The Peter Principle,* Peter and Hull (1969) coin "Peter's Placebo: an ounce of image is worth a pound of performance." Unfortunately, I've found this clever saying to be true. However, in the long run, competence or "performance" creates an important part of your image. Achieving strong business results, giving accurate advice, and being dependable are all part of a sound image. Beyond that, look to your role models for examples of how to create a good image in your organization. Other traits that stand out as a part of a sound image are

- seriousness towards your work but with an appropriate sense of humor and friendliness,
- mental sharpness, and
- a confident and relaxed nature that is not arrogant.

Behind your image lies your attitude. I promise not to give you a lecture about attitude. Instead I'll pass on an anonymous quotation I read recently on my calendar.

Attitude
The environment you fashion out of your thoughts...your beliefs...your ideals...your philosophy...is the only climate you will ever live in.

Probably your most influential exposure will come in meetings, especially those where you make a presentation. In these circumstances you usually have quite a bit of control. You can do whatever it takes to make sure your presentation is well organized, well presented, and that the message is on target.

Handling difficult situations is another chance for positive or negative exposure. Every situation will be different, but the most effective response will always be the same: keep your cool and think. Keep your own emotions in check and try to defuse the situation if that's appropriate. The listening skills we discussed earlier will help you sort out the various opinions and positions of the people engaged in the situation. You can then weigh all the aspects of the problem and draw your own conclusion. As a new employee, you may choose to steer clear if an argument or conflict erupts. My manager once told me, "Stay out of the tall grass when bull elephants are jousting."

DEALING WITH OFFICE POLITICS

If you are establishing patrons in the organization (especially your boss) and if you are getting the right kind of exposure, you will have a good base of power from which to deal with the slings and arrows of office politics. The first step is to be aware of what's going on. What are the currents moving in the organization? Who is aligned with whom? Which persons are adversaries? What is the management's position on this issue, and who supports or opposes it?

The organization will have a certain culture, and you will gradually learn what behavior is tolerated and what isn't. There will be certain patterns that people will follow as they deal with management, peers, and subordinates. Learn what these "rules of the road" are and observe them to avoid a serious crash. Perhaps you're a little intimidated or put off by those thoughts—conforming to the norms of the organization. However, let me assure you that you have already done this before. When you were in college and needed money did you know which parent to approach and which buttons to push to get what you needed? Or when you wrecked the family car, did you not try to break the news in the best possible way? Or did you sense that a sibling was treated differently from you? This analogy is in no way complete, but it serves to illustrate that you can be aware of what's going on around you, so that you can act accordingly.

If you have a good power base and if you know what's going on, you don't need to engage in negative politics. In the article I mentioned earlier, Patricia Young (2003) quotes from her interview with Colin Gautrey, the managing director of a British firm Politics at Work, who claims, "You can diffuse [politics] by naming it and bringing it into the open." If you have the trust of your boss, you can do this in a private conversation and get a reaction before doing anything publicly. Young (2003) also offers the following six-step survival plan:

1. Never react emotionally. You will never win. Practice calm responses to the situation. Anticipate what tactics the office manipulator will use.
2. Learn the rules. What is the other person's ulterior motive? What is driving this game?
3. Build allies. It may be other victims of the same office game, or it might have to be the human resources department. There is strength in numbers.
4. Learn to negotiate. In their book, *The Games Companies Play*, Gerry Griffin and Ciaran Parker (2004) say that part of the art of diplomacy "lies in you being able to horse-trade effectively." What do you have to barter with?
5. Knowledge is power. Enough said.
6. Intuition. Trust yourself. If you can feel the knives lining up behind you, you are probably right. Don't just sit there, go out and get a flak jacket or at least get out of the bull's-eye.

One final piece of advice for emphasis: don't engage in negative office politics yourself—period. Develop your patrons and allies, get the right exposure to management, take defensive measures as necessary, and you will be able to negotiate the minefield of office politics and advance your career.

SOCIAL SKILLS

To put it into perspective, social skills are an art. People can go to great lengths and expense to succeed through social contacts. For instance, a person may take up hunting because an important, up-and-coming executive on the Board of Directors is an avid hunter. He takes shooting lessons and buys expensive guns and gear. Eventually, he angles an invitation to go hunting with the executive. Others may join specific clubs or churches to seek social opportunities. Still others throw expensive dinner parties for influential people or plan exotic trips with them.

There isn't anything wrong with any of this, especially within the higher management levels. But this could be overdoing it for a project engineer. Nevertheless, you shouldn't avoid organizational social events or appear to be antisocial. Try to fit in. A working knowledge of table manners and social etiquette, combined with some good judgment, is all you need. And, of course, don't learn the hard way that drinking too much at an organizational function can be hazardous to your career.

One approach to social contacts is to treat them as any other type of networking contact. A social contact, even if it's just having lunch together, can provide a way of gaining rapport with someone, be they management, peers, or subordinates. For example, people who enjoy fishing or the theater or NASCAR racing can always steer the conversation around to that subject with someone who has that common interest. Such a conversation could eventually lead to an outing, an acquaintance, or a friendship. This approach, however, should evolve naturally and be handled with subtlety and tact. Take things at their natural pace.

I'll leave the rest of the social maneuvering to you. After all, it is an art.

PERSPECTIVE REVISITED

So that's Business World 101: competence, office politics, and social skills. By far the most important is competence. It's probably within the grasp of each of you to be a competent, hard-working project engineer with good business judgment. I've rolled a lot under the banner of competence, but with the right effort and training early in your career you can achieve it. However, to climb the ladder of success you must be able to get along with the people at work on a continuous and sustained basis. You should be aware of office politics and build a base of information power, expert power, referent (personality) power, and connection power from which to work. Spend some time establishing and maintaining your networks with management, peers, and subordinates.

Patrons are important to success. Patrons are management sponsors who respect your competence, opinions, and judgment. They can either be in your line of management (which is preferable) or in other branches of management (which is useful). For most engineers, career progress will be at an ordinary pace. For a few, it will accelerate in dramatic jumps. The patron gets ahead and pulls a few key engineers along. Patrons need people like you to give them the help and information they require to do their job.

Exposure is also important, since this is what builds the management's respect for your competence, opinions, and judgment. The foundation of competence and effectiveness must be in place for the exposure to get you ahead.

Social opportunities can be an important part of your networking and can be used with tact and subtlety, as they present themselves.

Above all, keep in touch with your basic life goals and aspirations. Even though it's important, your career is only one aspect of your life. Strive for excellence, like the street sweeper in Dr. King's quotation, but don't forget the ones you love. Cultivate those relationships. A friend once rhetorically asked me, "How many people on their death bed would say, 'I wish I had spent more time in the office'?" On the other hand, doing things in a success-oriented, personally effective way won't take any more time. In fact, it will probably save time because you're more efficient. Working smarter will build your self-esteem. It will enhance the way you feel towards your work. It will help you reach your potential on the job and attain your goals and aspirations.

REFERENCES

Griffin, G., and Parker, C., *The Games Companies Play: An Insider's Guide to Surviving and Winning in Office Politics* (John Wiley & Sons, Inc., 2004).

Hersey, P., and Blanchard, K. H., *Management of Organizational Behavior: Utilizing Human Resources* (4th ed.) (Prentice Hall, Inc., New Jersey, 1982), pp. 176–191.

Peter, L., and Hull, R., *The Peter Principle* (William Morrow & Company, Inc., New York, 1969), p. 165.

Petkovic, Ruzica A., PhD, personal communication, June 27, 2003.

Young, P., "When Good Isn't Good Enough," *The Globe and Mail* (Calgary, Canada, September 17, 2003), pp. C1, 6.

Chapter 7

Things That Can Get You Fired

When it comes to business ethics, your only viable strategy is obedience of laws, regulations, and important organizational ethics policies.

Unfortunately, this doesn't make a good story, so you won't find much validation of that strategy in TV drama, films, and fiction of all types. Even so-called reality shows tend to glamorize deception and callous disregard for straightforward, ethical behavior.

For a better reality check, you need go no farther than the headlines of any major newspaper or Internet news service. You'll see what happens when once powerful executives break the law or disregard the established principles of corporate governance. Take a recent headline, for instance:

Hard Landing for Skilling
Former Enron CEO Sentenced to 24 Years and 4 Months
Judge orders him to pay $45 million in restitution (Hays, 2006).

On TV newscasts you will see footage of CEOs or CFOs being led in handcuffs from their corporate offices. This is just the visible face of unethical, illegal corporate behavior.

The vast majority of people I've been associated with in business have integrity. But there are those who don't. It's not uncommon for significant-sized, established companies to have terminated a few personnel for illegal or unethical behavior. Exposing yourself to this consequence doesn't make sense, no matter how large the potential reward or how remote the chance of getting caught. The good reputation you've built over many years can disappear in an instant of bad judgment.

I invite you to give serious consideration to six areas of business ethics where bad judgment can be problematic and could get you fired:

- Laws and regulations
- False reporting
- Drugs and alcohol
- Harassment
- Conflict of interest
- Other organizational ethics policies

If you're savvy about these matters, rather than naïve or blasé, your chances of getting fired for them are remote. As a guiding principle, openness and honesty in your communication and intentions are the keys.

LAWS AND REGULATIONS

FINANCE AND ACCOUNTING

First let's enter the realm of the accountants. Let's talk about money. As a driving force, it is pervasive. It's the driver that has caused companies to bend the rules by such tactics as inflating earnings or presenting audited financial statements calculated to mislead the investment community. When money becomes more important to a company than its values, it is sliding down the slippery slope toward its own destruction.

The tension between achieving results and obeying the law is also present at the project engineer's level. Some temptations are clear and easy to recognize, and your actions are clear:

- Don't embezzle money from the company.
- Don't steal company property.
- Don't divert company resources of any kind for your personal gain.
- Don't lie or cheat on your travel expense reports.

Just because they're obvious doesn't mean that those temptations are easy to steer clear of. Take the last item above, for example. What if your boss indirectly hints that everyone puts a little extra on their travel expense reports? You find yourself traveling with the boss, and you're afraid if you don't pad your expense statement, you will expose your boss's illegal practices. What do you do? The answer is, "Don't lie or cheat on your travel expense reports." History abundantly demonstrates, and you yourself have probably learned a long time ago, that the "everybody's doing it" argument is a sure sign of trouble.

Don't fall into the trap of belittling established accounting practices. Here—like nowhere else—a *beat-the-system attitude* can get you into real trouble. Recognized accounting practices carry the force of law.

As a project engineer, you may find yourself in a position where you are required to sign for things. For example, you might be approving invoices for equipment deliveries or signing timesheets for employees or contractors. Talk to someone in your accounting group to learn what they require. If you have been granted approval authority for invoices, expense reports, timesheets, or other transactions, it's wise to carry out some *due diligence* each time you exercise that authority. Sometimes when you are signing invoices you are simply verifying that the goods have been received and are acceptable. Here your due diligence is to check the goods before sending the invoice to your boss for his or her approval.

If you're actually approving the invoice for payment, don't approve more than you are authorized to sign for, and don't use the *salami approach* that forces the supplier to divide large invoices into smaller invoices that fall within your financial authority. Say that you receive an invoice for $23,573 and your approval authority is $20,000. Also, the budget for this item is $19,000. It's wrong to ask the supplier to send two invoices that you can approve—one for $19,000 and one for $4,573. Instead, sign the original invoice and send it to your boss with a note explaining that the invoice is over your authority and exceeds the original budget by $4,573.

Commercial processes such as bidding, contracting, and purchasing have their own sets of laws and company rules to follow. Even the hint of improper commercial practices—like biased evaluation of bids—is a serious matter. Here you need the advice of an experienced and trusted person, such as a supervisor, team leader, or senior advisor, who can explain the detailed ground rules. These will vary from company to company.

Unfortunately for our peace of mind, the commercial side of the business world is full of *gray-area decisions*. An executive with a contracting firm once shared this observation with me: "People in the business world only do things they want to do and things they have to do. And no one wants to give you money." By implication, one has to put the other contractual parties in a position that forces them to pay.

Think about the situation Chet created in our case study in Chapter 5. He overstated the cost of changes caused by the client's decision to switch from a top-nozzle to a bottom-nozzle compressor. In doing so, he manipulated the contractor and the client into thinking that they had to pay millions of extra dollars and suffer a costly project delay. Were it not for Jeff's personal integrity, Chet would have probably gotten away with it. When faced with a choice between lying and having to resign his job, Jeff chose the high road. That's a tough decision, but one that may be necessary under extreme circumstances.

Others would argue that Chet was operating in the gray area—something that occurs all the time. Whether Chet violated the law would depend on what he said, what he wrote, what he knew, when he knew it, and his intent. After all, project changes often end up costing more than the amounts written initially in a change order. Perhaps Chet was just being smart and playing it on the safe side. In other words, unethical practices can usually be rationalized or explained away.

It's not easy, is it? The best you can do is

- evaluate the situation,
- discuss it with others and seek advice,
- try to isolate what—if anything—is illegal or unethical, and
- decide.

This transparent approach will keep you out of legal trouble. Remember, honesty and openness is the best policy.

ANTITRUST

Competition is the force that powers the world's economy and makes it thrive. It promotes efficiency and drives down prices. Without any type of regulation, history has shown that monopolistic business practices develop, which limit competition between private companies. On the other end of the scale, governments' attempts to run companies and industries, to avoid these private monopolies, have tended to be notoriously inefficient. Most governments have evolved towards policies that regulate business practices in order to promote and maintain competition, without interfering with pricing, production, or the other decisions of private companies, according to the Business for Social Responsibility website (2006) (a research and consulting organization on corporate social responsibility). Those governmental policies are implemented through antitrust laws and regulations.

If you find yourself dealing with *competitors* or other companies on your job, it would be wise to familiarize yourself with antitrust law. Your organization's lawyer can give you specific guidance on what you can and can't do or say. If you want some general information before going to the lawyer (or if your company is too small to have in-house counsel), the Federal Trade Commission's "A Plain English Guide to Antitrust Laws" (2007) is a good place to begin. It lists some of the main types of illegal, uncompetitive practices. Some of these are between competing companies (horizontal agreements between competitors):

- Price fixing
- Agreements to restrict output
- Boycotts
- Agreements to divide the market
- Agreements to restrict advertising
- Codes of ethics that restrict how professionals can compete

Other forbidden practices are vertical agreements, between buyers and sellers:

- Price fixing between suppliers and dealers that set minimum resale prices
- Nonprice agreements that restrict how or where dealers may sell
- Agreements that tie in the sale of one product with other products or services that the customer may not want

Working on industrial committees (e.g., ASCE or ASME) to standardize design codes sounds like a high-minded activity, but because it brings employees of competing companies together, those committees must be conducted with care. Likewise, joint industry projects financed by competitors to develop new technology must be structured with antitrust laws in mind. If you find yourself in a situation that brings you in contact with competitors and seems suspicious, take it up with your supervisor, your organization's legal counsel, or both. If in doubt, communicate. Your manager or your organization's legal counsel would both rather hear about a nonproblem than be blind-sided by one they had no idea was coming.

BRIBERY AND CORRUPTION

Probably the most obvious form of anticompetitive behavior is bribery and corruption. This is recognized as unethical in nearly every society around the world. Not only is it unethical, but it also

- disrupts competition,
- causes serious economical, political, and environmental damage, and
- is the single greatest impediment to poverty alleviation on this planet, according to the World Bank (Business for Social Responsibility, 2006).

The United States is a leader in the fight against international bribery and corruption. It enacted the Foreign Corrupt Practices Act (FCPA), which "makes it a federal offense to give anything of value to foreign officials, political parties or candidates for public office in order to obtain a contract or business relationship. The FCPA applies to U.S. citizens and companies around the world and to non-U.S. firms listed in the New York Stock Exchange" (Business for Social Responsibility, 2006).

If you work in business and are in any way involved with foreign governmental officials, you should become familiar with both the FCPA and your company's related policies. Things like paying for a foreign official's transportation, hotel rooms, meals, and other expenses may violate that statute or your company's policy. Certain "facilitating payments," such as fees to secure customs clearance, may be legal but are often subject to limitations. Don't fool around with payments to foreign government officials. *Get advice before you get involved.*

CLASSIFIED, PROPRIETARY, AND OTHER CONFIDENTIAL INFORMATION

Classified Governmental Information

If you're employed by a governmental organization in a position that has access to classified information, you'll require an appropriate security clearance. The laws and regulations for handling, securing, and disclosing classified information are strictly enforced. It's good sense—not to mention self-preservation—to become aware of those rules and obey them.

Proprietary Information

If you join a private sector company, you may be asked to sign an employment contract that restricts you from disclosing the company's proprietary information. Proprietary information includes technical information, trade secrets, computer programs, manufacturing processes, formulas, sensitive data, customer lists, business plans, and strategies—anything that gives the company a competitive edge in the market place and is considered by them as proprietary. Federal laws prohibit the unauthorized release and use of this information, and it's undoubtedly clear that you shouldn't do that. It would harm your own company.

It's equally unlawful to use other companies' proprietary information. Say, for example, that you work for one company and then move to another. You should not take any proprietary information with you or use it in your new company. You could get yourself and your new employer in serious trouble. Certainly, this doesn't apply to *everything* you know or learned with the first employer. You will have to use your best judgment to navigate through this one.

Sometimes other companies will offer proprietary drawings, data, or documents to you across the table in a meeting. *Unless your company is working with that company and has an established contract that covers the exchange of information, you should refuse to accept the proprietary material.* If you anticipate that this could happen before the meeting, you can get your legal department to prepare a confidentiality agreement or a nonconfidential disclosure agreement—whichever they deem best. A nonconfidential disclosure agreement will steer the agenda away from any proprietary information, since the other company agrees not to disclose their confidential information. Said another way, any information exchanged will be considered to be nonconfidential.

Let's clarify this because it's important and happens frequently. Consider a situation where you work for an established chemical company. It's a large company with a research subsidiary at another location. You are approached by a small start-up company that has an idea for a chemical process that could improve the production in the plant where you are currently assigned. They claim to have filed for a patent, which hasn't been issued. In this case it's clear that you should discuss the meeting with your legal department and probably have the other company sign a nonconfidential disclosure statement before holding the meeting. In the meantime your company's lawyer can check with the legal counsel at the research subsidiary to make sure your company is not working on a similar idea. If they are, you should probably cancel the meeting.

Without this formality, acceptance of proprietary information from the small start-up company could jeopardize your company's patent position or expose them to subsequent legal claims.

Similarly, if you receive proprietary information in the mail that is not transmitted under an existing agreement, immediately send it to your legal department.

Other Confidential Information

As a new employee in the private sector, you'll undoubtedly come across other types of confidential information. Even though not all of it is protected by laws and regulations, it's covered in this section for completeness. As a matter of good practice and integrity, this information is disclosed on a need-to-know basis.

Commercial and technical bidding information must be strictly safeguarded. There are illegal international entities that engage in information brokering for profit.

Information relating to personnel matters—such as salary, promotions, bonuses, disciplinary actions, transfers, employee confidential data, and records—is often stamped "confidential" or "private." It's sensible to always handle that information with discretion and respect for people's privacy. The gossip unleashed by careless or malicious release of confidential personal information can hurt both individuals and the company.

Still other confidential information is considered so sensitive that its disclosure is formally *restricted to specific company personnel*. Records are kept of the distribution. For example, documents relating to mergers, acquisitions, business deals, joint ventures, stock issues, layoffs, or plant closures would normally be restricted to prevent premature release. The use of this or any other type of insider information for personal gain is discussed in the "Conflict of Interest" section, later in this chapter. For now I'll just say: don't violate either the letter or the spirit of a restriction against disclosure to other than authorized personnel that need to know it.

FALSE REPORTING

All organizations must rely on their employees and officers to accurately record business information. They use this information to make decisions, file accurate financial and tax reports, and conduct business of all types. If a culture of dishonesty and false reporting develops, the whole organization's reputation can become tainted. It's like a drop of ink in a glass of water. It spreads everywhere.

EMPLOYMENT APPLICATION

Your attitude toward reporting information starts with filling out an employment application. You should put your best foot forward when selling yourself but only submit accurate information. Enough said.

RECORDS AND TIMESHEETS

Accurate reporting is a value and a priority. Consider a new employee who hasn't yet accumulated any vacation or time off and who has been working long hours for over 6 months. Her spouse's parents have come from overseas for a short visit in the middle of the week, and she would like to spend some time with them before they return home. She and her spouse decide to go to the beach.

How should she handle this? She could probably just enter eight hours of sick time and get away with it, but that would be false reporting. Instead, she decides to ask her boss's permission. He says, "OK," but offers no advice on how to report the time. When it comes time to fill out her timesheet, she can't enter vacation because she has none. Probably her best course of action is to enter the day as "time off with pay." Her boss will review the timesheet when he approves it. She always risks the possibility that she may not be paid for that day, but in the vast scheme of things that's better than false reporting.

TEST RESULTS, DATA, AND RESEARCH RESULTS

Engineers often find themselves collecting data and conducting tests to make sure something will work like it's supposed to. People depend on those results for their personal safety. The operational integrity of processes, plants, and factories can depend on reliable pilot-scale test data. Product testing assures the safety, health, and satisfaction of consumers. Environmental testing bears on the health, safety, and quality-of-life for whole communities. The structural integrity of buildings, bridges, dams, ships, vehicles, aircraft, pipelines, and other industrial structures is a significant public safety issue.

It's obvious that test results must be reported accurately because people depend on them. Test results measure and assure quality. They ensure safety and health. They underpin the economic viability of new ventures. But an engineer can feel the pressure to compromise his or her integrity when, in the lab or out in the field, the data don't agree with the theory or don't produce the desired response. *In situations like this, an engineer must resolve to report what is observed, no matter what.* People can, to the best of their ability, explain any discrepancies that occur and seek expert advice to interpret them—but it's wrong to fudge the data. This is where speaking and writing skills can serve a person well. Keep the management informed, and if someone tries to cover things up, press for open and accurate reporting. It's hard for anyone to argue against that.

REQUIRED REPORTING

Despite our best efforts, incidents happen. Accidents occur and people get hurt. Noxious fumes are emitted into the air. Chemicals are spilled into bodies of water. Contaminants leak into the soil. The ramifications of these situations are never pleasant to accept, especially if it occurs on your watch and you are responsible. *But cover-up is never a good option.* The best action one can take is to immediately respond by calling response teams, helping the injured, and controlling the extent of the damage. Then the incident should be reported, as soon as possible, according to your organization's procedures, even if it means calling the supervisor halfway around the world in the middle of the night.

DRUGS AND ALCOHOL

Most organizations have no tolerance for drug and alcohol use on the job. It exposes them to considerable liability. If, for example, a drunken employee causes death, injury, or considerable property damage to a third party, the company would probably be liable. If, in addition, they have no drug and alcohol policy or haven't enforced the one they have, their liability would be greater.

Drug testing is often a part of the hiring process. Once hired, an employee in a position where sobriety is critical to his or her ability to perform may be subjected to random drug testing. Most people with whom I've associated in recent years won't drink any alcohol over lunch to stay in compliance with their company's drug and alcohol policy. The days of the three-martini lunch are definitely gone in many companies.

Even if your company doesn't have a drug and alcohol policy, you still must be careful. More and more, companies are putting clauses in their contracts requiring that

- contractors have a drug and alcohol policy, and
- work done on the client's property is done in accordance with the client's drug and alcohol policy.

A contractor's employee who has a few beers over lunch and returns to the client's facilities to continue working could be in violation of the client's alcohol policy. It's best to wait until happy hour. Coming to work under the influence is one of the surest ways to get fired.

HARASSMENT

Well-managed companies recognize that all workers should be treated with dignity and respect. Those companies know the devastating and debilitating consequences of harassing behavior in the workplace. Harassment exposes companies to serious legal liability and tears at the moral fabric of employee teamwork and performance.

Federal discrimination laws prohibit harassment based on race, color, sex, age, religion, physical or mental disability, national origin, citizenship status, and veteran status. Those kinds of harassment, by one employee towards another, can subject the employer to liability. If the harassing employee is a supervisor or manager, the employer is more deeply implicated. Laws in some states hold both employers and individual employees liable for harassing incidents.

The real damage of harassment is more subtle. In Chapter 3, we considered the people-oriented aspects of effective management:

- Motivate
- Build a team
- Develop people

Persistent harassment in a work place creates exactly the opposite effect. The harassed employees become unmotivated and stop looking for ways to contribute. The team is fragmented into individuals—or even worse, cliques that aggressively oppose each other. Employees' performance degrades rather than develops.

What is harassment? Behavior by an individual or group that another individual or group finds intimidating, hostile, degrading, or offensive constitutes harassment. It's generally judged to be harassing behavior by the person or persons who are the objective of the remarks or actions. The legal standard is whether a reasonable person would be offended or threatened. Let's consider some examples that are usually harassment:

- Profanity, particularly the vulgar type
- Ethnic jokes or slurs
- Racial jokes or slurs
- Comments about someone's sex (salty operations manager repeatedly refers to a woman employee as "that girl" or uses an even more insulting synonym)
- Solicitation of sexual favors in exchange for favored treatment or career advancements
- Jokes, pin-up posters, email, notes, conversations, telephone calls, or other means that convey sexual connotations or innuendo
- Unwanted touching or sexual advances
- Mimicking a disabled person's movements or behavior

- Derogatory remarks or innuendo about someone's sexual preference
- Bullying, threatening, verbal abuse, or cruelty
- And so on

Some related situations don't necessarily constitute harassment. For example, an employee may feel intimidated by a supervisor's performance evaluation, but the supervisor is within his or her right to do this type of counseling. If, however, the supervisor threatens or verbally abuses the employee, the situation could arguably be harassment. There is a fine line here, where doubt often falls in favor of the supervisor.

The area of sexual harassment deserves some additional comments. There is nothing wrong with people meeting in the work place and initiating a relationship. It is the unwanted advances that are considered harassment. It's also possible for a supervisor and an employee to have a relationship. However, one of them should be reassigned because of the obvious conflicts of interest regarding promotions, pay increases, and other personnel actions.

Your company or organization will likely have a policy concerning harassment. Even if they don't, you should be aware of these basics and act accordingly. The laws demand it.

If you are the victim or witness of harassment, it's best to handle it according to company policy. The first step is usually to confront the harasser to serve notice that the behavior is unwanted or considered inappropriate. In many cases that will stop the harassment. The next step is often to report the incident to your supervisor or the Personnel or Human Relations Department. This requires a lot of judgment, so seek advice from trusted friends.

CONFLICT OF INTEREST

Conflict of interest is often one of the most difficult business ethics issues for inexperienced employees to discern. Most companies and other organizations have conflict-of-interest policies that prohibit employees from entering into situations where their own interests are contrary to the interests of the organization. In most cases, the policy covers the employee, as well as members of the employee's immediate family. Both *actual* and *apparent* conflicts of interest generally fall under the policy, which means that even the appearance of inappropriate activity must be avoided.

Let's suppose that a graduate student is employed on a government research project while he is still in school. The project is successful, and the governmental agency decides to competitively bid contracts to build several large facilities. Because of his key role in the research work, the student has been assigned to a team that will evaluate the technical portion of the bid proposals. Just prior to the bidding, an employee of one of the bidders contacts the student. He tells the student that they have been impressed with his work and will offer the student a job if they are awarded the contract. Is this a conflict of interest?

In this case, the governmental agency's interest is to achieve a fair and objective evaluation of each bidder's technical proposal so that they can determine if those proposals are technically acceptable. The student's professional interest is to do a good job for the government, but his personal interests include getting hired in the spring when he graduates. The conditional job offer could potentially influence his decisions while analyzing the bids. This student *has* a conflict of interest.

The ethical way to approach conflict-of-interest situations is with openness toward the organization's management. In this hypothetical situation, the graduate student should discuss the matter with his immediate supervisor on the government's bid evaluation team. There will also likely be conversations between that supervisor and his or her management until the matter is resolved. In this instance, conflict of interest is not the only ethical issue. The contractor's inappropriate employment offer is much more serious and must also be dealt with.

Another important consideration, from the student's point of view, is documentation. The student should keep any email or memos that he has sent or received. If everything was done by word of mouth, the student could write a memo to the supervisor, documenting his understanding of what happened and what was decided. Of course, he should keep a copy. Openness with his management will help keep this employee out of trouble.

TYPICAL CONFLICTS OF INTEREST

Conflicts of interest can take on many different forms. One of the most obvious is for the employee or an immediate family member to hold a financial interest in a company that is selling goods or services to the employer. Another closely related conflict occurs when an employee or family member has a financial interest in a company that competes with the employer. Owning publicly traded stock in a competing company probably will not be considered a conflict unless insider information was involved in the purchase or sale of that stock.

A more subtle conflict of interest occurs when an employee receives personal gain by virtue of his or her position in an organization. Expensive gifts, trips, or entertainment provided by suppliers should be avoided. For example, accepting a $30 pen from a contractor at a contract-signing ceremony is probably permissible, while a $500 pen and pencil set is not. However, promotional items (with a logo on them) of nominal value are generally (but not always) OK. Cash or its equivalent is another matter. Special discounts, free services, cash, gift cards, stock, or anything similar received from a supplier could be considered a kickback.

Federal government employees can accept virtually nothing from contractors—not even a meal. (I'm reluctant to go into any detail on the conflict of interest or other policies and ethical guidelines for federal employees. Let it suffice to say that the rules are strict, and if you work for the federal government, you need to know them.)

The use of information that is not generally available to the public can also represent a conflict of interest. Buying or selling stock based on specific, insider information is a violation of most business conduct policies and is most likely illegal.

Spouses who work at similar jobs in competing companies and have access to proprietary business or technical information could have conflicts of interests. Consider a woman recently promoted from the research staff of a major drug manufacturer to a corporate planning position. Her husband works as a sales manager for a competing drug firm. In this case, both of them should take the situation up with their respective managements to seek a solution that all parties can accept. Of course, they should each document the matter and the decision by their respective companies and keep copies in their personal files.

AWARENESS

This should give you a cursory understanding of what conflicts of interest are and how to handle them. The difficult part is usually recognizing that you are in a conflict-of-interest situation. Your organization may have a detailed description of the policy with examples. If not, you are on your own. As a last resort, visualize how the situation would come across on the front page of the local newspaper— for example, "Project Engineer Accepts Hunting Trip Prior to Bid Opening."

Once you recognize a conflict-of-interest situation, take it up with your supervisor and attempt to keep some documentation for your personal files.

OTHER ORGANIZATIONAL ETHICS POLICIES

Some companies and other organizations have isolated policies that may seem to be more strictly enforced than they should be. These policies may have resulted from a serious incident in the organization's past or merely reflect a strong belief by the management. Take, for example, the use of a company credit card for personal expenses. Some companies may allow it, while others will terminate an employee who uses the card to go on a ski vacation or buy personal clothing.

The same can apply to the personal use of the organization's Internet access, email, copy machines, long-distance phone service, and so forth. Many organizations tolerate insignificant usage but prohibit abuses. Others have a zero tolerance policy to some or all of those activities. Almost all organizations prohibit the use of the company's or government's property for personal gain.

As a new employee, become aware of the policies that are strictly enforced before you accidentally kick a hornets' nest of management attention for some act that you considered trivial. Your fellow employees and certainly your supervisor will know what those zero tolerance policies are.

DRAWING THE LINE

We live in a world of practicalities, and it's sometimes difficult to discern the line between ethical and unethical behavior. You are pressed from all sides to deliver results, but there may be regulations or policies that occasionally stand in your way. If you find yourself in a predicament like this, get your facts straight and carefully analyze the situation. As you consider your possible courses of action, weigh the consequences of each. Try to sort out what the real ethical issues are and then tend to them. The starting point is clear—don't break laws or regulations and don't lie, cheat, or steal. Those values are the foundation of your integrity.

But there may be times when things aren't clear, and you will need the advice of a trusted mentor, supervisor, or other experienced person. You could be exaggerating or misunderstanding something that really isn't a problem. Maintaining openness, especially with your supervisor or management, is one of your most important tools for dealing with conflicts of interest and other ethical situations where the correct choices aren't always clear or depend on information you don't know. If the situation has serious legal implications, you may have to seek advice from your organization's legal counsel. In very extreme circumstances you might even have to seek personal legal counsel outside of your organization.

From an ethical point of view, the first few years of your career are important. They give you the ethical ground truth you will use to navigate through your encounters in the workplace for the rest of your career. If your organization has high ethical standards, you will be well equipped. If not, I hope you are able to recognize, as Jeff did in our case study in Chapter 5, that it's time to change jobs.

If you still aren't convinced that ethical behavior is the right course for you, read the best-selling book *The Integrity Advantage*, by Adrian Gostick and Dana Telford (2003). It builds a convincing case—based on the advice of business leaders—that integrity gives you a competitive advantage in the business world. That's right—ethically sound conduct is good business!

Business ethics are something to be aware of and not something to be afraid of. They're part of your integrity. That integrity will help you make sound business decisions in those difficult situations where it's tough to draw the line. As I said before:

- Know the relevant laws, regulations, and your organization's business ethics policies.
- Evaluate the situation.
- Discuss it with others and seek advice.
- Try to isolate what's illegal or unethical and avoid it.
- Treat the situation with openness and honesty.
- Then decide.

To help you decide, you can always use the newspaper headline test mentioned earlier. If your actions couldn't stand up to that test, change your approach. By conducting your business in this way, you will also gain respect, which will help you rise in the organization as a leader.

REFERENCES

Business for Social Responsibility, "Overview of Business Ethics," http://www.bsr.org/CSRResources/IssueBriefDetail.cfm?DocumentID=48815 (2006).

Federal Trade Commission, "A Plain English Guide to Antitrust Laws," http://www.ftc.gov/bc/compguide/index.htm (2007).

Gostick, A., and Telford, D., *The Integrity Advantage* (Gibbs Smith, Publisher, Salt Lake City, 2003).

Hays, K., "Hard Landing for Skilling," *Houston Chronicle* (Houston, Texas, Vol. 106, No. 11, October 24, 2006), p. A1.

Chapter 8

International Business Skills

Your culture forms your values.
Culture tells you what's right and wrong.
It tells you what you like and don't like.

Anonymous

THE CULTURAL GAME

The crisp mountain air filled the lungs of the up-and-coming management students as they walked briskly from the lunch room to their afternoon class. Low sun gilded the brown leaves strewn on the sidewalk. "What do we have this afternoon?" Jack asked.

"I don't know," Jillian said, pulling her jacket around her shoulders. "I heard it would be some kind of group dynamics exercise."

As they entered the meeting room, their instructor assigned Jack and Jillian to a group. "Don't talk to anyone until you come back to this room!" he ordered. "Go down this hall to Room D and take a seat at the table."

"OK," Jack said.

"No talking!"

A single table with a dark tablecloth and four chairs stood in the center of the small break-out room. Already seated across from each other were two class-mates—the opposition. A deck of cards was stacked in the center of the table and a sheet of instructions lay face down in front of each player. Jillian took the chair marked dealer, while Jack sat across from her.

Another instructor came through the door. "Read the instructions and begin when you're ready. You'll play five tricks. The winners will move to Room E and the others remain here...and remember, no talking."

Jack looked at the rules as Jillian began to deal. King is high...Ace is low.

Jack and Jillian won the first two tricks, lost the third, and then won the last two. *We're on a roll* Jack thought. He smiled with satisfaction at Jillian as they got up and moved to Room E.

The room was equipped the same as the other, except there were no instruction sheets. Two unhappy classmates who had lost the previous round sat across from each other.

The instructor gave them a nod and one of the others dealt the hands. The players tossed King, Queen, Ace and Ten in the center of the table. Two hands shot out to claim the trick. There was a moment of awkwardness as Jack and the burly fellow beside him both held onto the cards. Jack conceded.

On the next trick the same thing happened. This time Jack wrenched the cards away and stacked them beside his place. The other team looked at the instructor but got no response. The situation repeated itself three more times with the anger and frustration mounting.

Jillian and Jack lost a game they felt they should have won. The winners left the room. Jack and Jillian remained behind, feeling branded as losers. In came another pair of opponents, still there were no instructions, and the process continued once more with the same hostile results—only worse.

In the wrap-up session that followed the third game, the point of the exercise was indelibly etched on Jillian's, Jack's, and everyone else's memory. An after-lunch card game had woken the players to the basic dilemma of working with people from other cultures. Though those participants were not from different cultures, the factors that make international communication difficult were wrapped up in this card game:

- It was difficult or nearly impossible to communicate effectively and solve even the most obvious problems.
- No one *really* knew the rules of the game.
- Eventually, everyone lost when tempers flared and polite behavior spun out of control.

GLOBAL BUSINESS

In today's business world, it's essential to know how to work effectively with people from other cultures. To remain competitive, companies seek cost-effective labor sources, which take them to China, Southeast Asia, India, and Mexico, to name a few. The search for minerals such as oil and gas propels companies to all corners of the globe—Africa, Asia, Australia, the Arctic, Europe, the Middle East, and Russia, not to mention North, Central, and South America. To sell their products, companies must engage global markets, while making their products attractive to the new markets or breaking down the trade barriers that exist.

And companies with a presence abroad can't just drop in every once in a while to see how things are going. There are deals to make, agreements to negotiate, factories to build, and workers to hire, train, and motivate. Raw materials, equipment, and other resources must be purchased and shipped to the worksites. Products must be produced, sold, and transported to markets. There are quality or other production problems to be solved and delivery schedules to be met. There are safety issues to be dealt with, labor disputes to resolve, and local politics to contend with. The company presence must be built up and management teams assembled from foreign and local sources. Costs must be kept under control for the business enterprise to be successful. Business controls, accounting systems, banking—all must function with integrity and efficiency. It takes a presence and a thorough understanding of the host country and its people to make all this happen.

And once again there are project engineers at the workface taking responsibility for their areas: coordinating, integrating, asking questions, solving problems, and tending to the people side of their business. Except this time the people think and act in ways that don't seem to make sense.

Whether you live and work in a foreign country or travel internationally on business, being effective (*how you do your job*) is an essential tool in your tool kit. *Success depends on being able to adapt to your new surroundings and get along with the people.* Sound familiar? It should—except working abroad intensifies the challenge by an order of magnitude.

START WITH YOURSELF

When you move to another country you should start by taking care of yourself. There are strong disruptive forces working in your life that you should not be taken lightly. You have a job to do, and you're excited and anxious to get on with it. But you also have some adjustments to make before you can function effectively in this new culture.

CULTURE SHOCK IS REAL

From the moment you arrive you see the differences. The buildings look different. Cars and trucks are different and might be driving on the *wrong side of the road*. The people may look unusual and dress oddly to your eyes as they crowd onto subways or buses and fill the teeming market places. If it's a developing country, you will undoubtedly see large segments of the population living in abject poverty—sorting through garbage to find something of value. The flood of new impulses will interest, repulse, delight, and shock you.

At the same time you are experiencing all the differences, you may be alone. You've been separated from your friends, loved ones, and the rest of your normal support system. Sure, you have email, but it's not the same. You begin to grieve the separation from friends and family.

If the family accompanies you, their adjustment may be more difficult than yours. They feel the same separation from friends, playmates, schoolmates, and other support, but they don't have the challenge of working to occupy their thoughts. I recently spoke to an expatriate wife whose husband was working in the Chinese oil business around Bo Hai Bay. She passionately expressed her frustration and the isolation of living in a small town, unable to communicate with her neighbors, and with no Western friends.

At work almost all the faces are new to you. You may know one or two coworkers from a previous assignment, but nearly everyone else, both expatriates and host country nationals, are unfamiliar. You try to get traction on the new job but you don't know

- whom to contact for necessary services,
- how the organization works,
- the rules of behavior,
- whom to trust,
- and the list goes on.

Eventually, you will learn all that—just like you always have—but in this setting, things seem more threatening. Your repertoire of heretofore successful tactics don't seem to get the expected response, especially from the host country people. You may begin to feel your effectiveness isn't what it was on your previous assignment and certainly not what it should be. Over the weeks, your self-confidence and self-esteem may begin to erode. You're in shock—culture shock.

When you move to a new culture, your first order of business is, as usual, to take care of the basics of life: food, shelter, and security. But it's equally critical to recognize that you (and your family if they are with you) are suffering from culture shock. It's virtually impossible for anyone—even the most experienced international people—to avoid it because it's "rooted in our psychological processes." Lane, DiStefano, and Maznevski (2006) further point out that "everyone experiences disorientation when entering another culture. The normal *assumptions* used by managers in their home cultures to *interpret perceptions* and to *communicate intentions,* no longer work in the new cultural environment." People become frustrated and feel ineffective when their normal attempts to socialize or conduct business result in confusion, frustration, and stress.

Most people adapt and work their way through this in the first year, depending on the situation and the person. "A smaller percentage either 'go native,' which is usually not an effective strategy, or experience very severe symptoms of inability to adjust (alcoholism, nervous breakdown, and so on).... A frequently reported explanation for expatriate failure has been poor adjustment of spouses" (Lane, DiStefano, and Maznevski, 2006). It's crucial to recognize culture shock and to cope with it, for your sake and for the sake of your family members, if they're in this together with you.

COPING STRATEGIES

A practical approach to coping with the new environment is to first fix the things that are bothering you. Rebuilding a support group and beginning to enjoy life in the new environment are high on the priority list. Also a high priority is to promptly learn what behaviors the people there find offensive and, of course, edit them out of your demeanor. Then you can deal with the heart of the matter—gaining the insights into the new culture, so you can adapt to living there and work effectively with the people. It's not always easy, but approaching the situation with an attitude of respect and collaboration will unlock the door.

New Support Group

Finding a new acquaintance in a foreign country may not be as difficult as you think. A natural place to look is among other expatriates, especially from your own organization or from your own country. Expatriates—whether in business, the military, or government service—recognize the transient nature of the situation and the importance of establishing new relationships. They reach out to newcomers and are generally quicker to form friendships than they would be in their home country. I'll give you an example. When I went to Norway as a bachelor, I was overwhelmed the day after my shipment of household goods arrived. That first morning I found the coffee pot that I had bought locally but couldn't find a cereal bowl and spoon. I went through several boxes, and the best I could come up with was two whiskey glasses. I filled one with milk and the other with cereal, and then alternated a slug from each until my essential hunger was satisfied. At work my boss asked how the move went. I related the story about my breakfast, and we had a good laugh over it. That evening, however, he and his wife showed up unannounced at my door with a pizza. They pitched in and helped put the kitchen and much of the rest of the living area in order. Their gesture let me know they cared about how I was doing beyond the boundaries of the job.

In well-established expatriate communities, organized clubs (like the American Club or the International Club) operate as a social hub with a restaurant catering to expatriate tastes, a pool and athletic facilities, social events, and a host of other activities. Informal women's clubs, organized at the company, governmental agency, and community level, provide important welcoming and social opportunities. Welcome packets and cookbooks developed by these organizations help families adapt to their new environment. International schools provide an important center of social activities for families with children. In addition to a wealth of local activities, these schools offer educational field trips, trips to other international schools in the region, and vacation trip opportunities. International church congregations provide similar opportunities. The people who take part in these organizations—young and adult—enrich their lives and build bonds of friendship that will last a lifetime. A friend can make a world of difference.

However, there are pitfalls when socializing with expatriates. It's easy to fall into the trap of finding fault with the new environment and comparing it to how good things were in the home country. Over time this can lead a person to lose respect for the host country people and their ways. It's good practice to tactfully avoid conversations that criticize the new country. Maintaining an attitude of respect is an important part of the adapting process, but it's absolutely essential to being effective in your job if you have any contact at all with the local population. We will go into this later in the chapter.

Nevertheless, being a part of the expatriate community is a source of social support and business contacts. And inevitably you will run across long-term residents who know and love the local culture and can give you glimpses of what it is really like.

Enriching Your Life

Enjoying yourself is something you don't want to give up when you move to a foreign country. The country itself will invariably have things to see and do. Be a tourist! Get books. Check the Internet. Take tours and talk to the tour guides. See churches and temples. Visit the tourist attractions. Experience the cuisine. Go to the tourist beaches. Visit the market place. Drive through the countryside (where it's safe). Talk to the people you meet when you can. Make your stay an adventure.

To enrich your day-to-day life you need insights into the world around you, which tend to be blocked by the language barrier. Language training is worth the intellectual effort and expense. It's a tall order to learn the language well enough to use it in business, but just being able to read the headlines of the local newspaper will open up new understanding. The ability to read labels on groceries, instructions on products, and signs on stores and public places will remove stress and frustration. People who just jump into the language and aren't deterred by their stupid mistakes seem to do the best. Their Southern drawl may sound a little strange in French, but they ignore that. Eventually they gain enough competence to read and understand, which makes their life less frustrating and more enjoyable. And there are those with linguistic ability who become fluent over the course of their assignment.

These two simple activities—being an adventuresome tourist and learning the language—will almost certainly add quality to your life and help you get through your culture shock.

Quickly Learning the Things to Avoid

We have already covered the primary thing to avoid—criticizing the new culture and its people. It's an attitude that is natural to have. If you're in culture shock, you're not feeling particularly good about yourself, and that spills over into your feelings toward others. It's also contagious when you're in an expatriate social setting with others who are adjusting to their new home. But criticizing the new culture and its people undermines your respect for them and sets you up for social and even professional failure.

Next in importance among the things to avoid, let me plant a warning sign in the ground about using humor. You'll find that getting the punch line of local jokes is one of the last language skills you will develop. Believe me when I say that it takes even longer to get the knack of telling funny, appropriate jokes. At best, your home country humor may be unfunny or irrelevant to your host country associates. At worst you'll discover the short circuit between careless humor and a personal, national, or even religious insult. Take it slowly, and avoid using humor until you really know what's funny and what's socially acceptable.

The other obvious things to avoid are those glaring social taboos and breaches of business etiquette that every culture has. For example, in Japan there seems to be a ritual around giving and receiving business cards. Each card is to be received graciously and read. To deface, mutilate, or treat the card with carelessness or indifference in the presence of its owner can give great offense. In Muslim cultures, showing the soles of your feet or touching the top of someone's head is considered rude. Even close cultures such as the United States, Britain, and Canada have their subtle differences that can lead to embarrassment, irritation, or friction.

Organizations that are culturally aware will include a list of behaviors to avoid in their orientation training. If you don't get this information shortly after you arrive, ask your boss or someone in the office. If necessary, compile your own list. It will keep you from making serious mistakes that can get you off on the wrong foot. As they say, you don't get a second chance to make a first impression.

Gaining Insights into the Culture

So far we've dealt with the remedies for culture shock and only hinted at how to be effective in your job. One piece of the effectiveness puzzle is gaining insights into how the host country people think. We'll use the basic communication model introduced in Chapter 6 as a starting point.

When you are trying to get your point across, you choose words from your own language that are based on your background and experience, which includes your culture. Even your body language, which to a certain extent comes from culture, conveys meaning. When the receiver is from another culture, he or she tries to interpret what you mean. Your slang or idioms are sure to be misinterpreted by all but the most experienced foreign listener. Even basic messages can be obscured or misinterpreted. Conversations move quickly, and it's impossible to analyze each word and phrase. Instead, we all have our *assumptions* about what certain words, inflections, phrases, and gestures mean. The listener instantly compares what he or she sees and hears with culturally derived assumptions and draws conclusions (Lane, DiStefano, and Maznevski, 2006). If you want to convey your intentions clearly, you have to understand the listener's assumptions so that you can phrase the message in the most effective manner. Likewise, knowing the foreign speaker's assumptions is the key to understanding her or his message.

To gain those insights into a foreign culture, it's essential to get to know the people. We've already mentioned the benefit of having conversations with long-term expatriate residents who understand the local ways. These conversations are a good source of general information but can be biased. For example, that expatriate may have had a bad experience with a certain business person or governmental official that has caused him to mistrust the local business community or government. He may be painting everyone with the same brush. There's no substitute for getting to know the host country people you are working with. Particularly in the Far East, the trust that builds through long-term association is an essential step in being accepted as a business associate. Those long-term relationships take nurturing, a lot of quality conversations, and perhaps drinking a lot of tea. Insights will come to both of you, helping to solve the intercultural communication puzzle. And the relationship will help you both weather the misunderstandings and business disagreements that are sure to occur eventually.

INTERCULTURAL SKILLS

Most of us who work abroad are thrown into it on rather short notice with little training or preparation. We find ourselves moving to a new assignment in Europe with less than a month's notice, or flying to West Africa on a business trip to present a contract award recommendation to governmental officials. In a matter of days we are up to our necks in cultural matters and need the skills to navigate through them. With no time to prepare, it is best that our tool kit contains

- the skills to help us communicate effectively across cultural lines, and
- a simple but versatile approach to decision making and problem solving that works in diverse settings.

It's also useful to have a framework to help identify and map cultural differences. Let's take these one at a time.

What Has to Change in the International Setting?

The short answer to this question is, "Not much." Fortunately, the fundamentals of competence—technical skills, hard work and business judgment—described in Chapter 6 are relevant to the international environment. Your technical and business competences are why you were selected for a foreign assignment and will be essential as the project unfolds. When you get into the job, you will continue to build your technical and business skills as you gain international experience.

The approaches for dealing with office politics are still applicable, and you will need to cultivate patrons. It will, however, take some time to get to know the other expatriates and to understand how to work with the host country employees in the office. Also, in the foreign setting new political forces can disrupt your work. Local events, labor disputes, and national politics can have immediate and unexpected impacts on your project. Your organization may have senior people monitoring those situations and providing timely information, warnings, and guidance to deal with these events, but small companies may not have that luxury. Over time, you'll learn the hot buttons and how to avoid them.

Your social behavior will gradually conform to local customs, and you should seek advice to avoid socially embarrassing situations. Take it slowly and learn what's politically correct from the other expatriates and your host country friends.

The major change will come on the people side of your job. If you're going to be effective, you'll have to quickly learn and use a set of cross-cultural communications skills to supplement the skills discussed in the "Personal Efficiency and Effectiveness" section of Chapter 6.

CROSS-CULTURAL COMMUNICATIONS SKILLS

Let's ask our basic question from Chapter 6 in a more international way: *"What does it take to get ahead when working with people from another culture?"*

Success hinges on effective communication and mutual understanding, but we know from Chapter 6 that communication can be difficult even between people of the same culture. In an earlier edition of their book *International Management Behavior*, Lane, DiStefano, and Maznevski (1997) present eight cross-cultural communication skills that should be added to the project engineer's repertoire to help overcome these difficulties, especially in the early stages of international relationships. These basic principles are applicable to effective communication anywhere in the world:

1. The capacity to communicate respect for other people's ways, their country, and their values
2. The capacity to accept the relativity of one's own knowledge and perceptions (i.e., accepting that an individual's knowledge, perceptions, and views are not universally regarded by others as being valid)
3. The capacity to be nonjudgmental
4. A tolerance for ambiguity

5. The capacity to display empathy
6. The capacity to be flexible
7. The capacity for turn taking
8. The humility to acknowledge what one doesn't know

The first one—the capacity to communicate respect—seems to be a prerequisite for fruitful international cooperation. If you can get across respect, the rest will usually follow. Number four, a tolerance for ambiguity, is particularly critical in an international setting. In the fog of cross-cultural communication, it's often difficult to discern the best course of action, let alone reach agreement on it. One has to get comfortable with proceeding under uncertain circumstances and hedging against the risks until the situation plays out and the fog clears a little. Sometimes the best thing to do is nothing, until better understanding exists between the parties. However, to the hard-charging North American project engineer and his or her management, "nothing" may be a very undesirable option.

The rest of the communication skills are self-explanatory, and I'll leave it to you to interpret them in the situations you encounter. Used together, those skills create the right climate for establishing trust and promoting the type of dialog that can avoid conflict and lead to solutions. I suggest you write the cross-cultural communication skills on a sheet of paper and keep them in front of you as a reminder when you're meeting with your foreign counterparts.

TIME, GOALS, AND PATIENCE

While visiting Southeast Asia on business, I took a weekend off to visit some ancient ruins. It was the off-season, and it turned out that I was the only tourist in my group. As I toured the grounds, I struck up a conversation with the guide, an insightful elderly woman. Somehow our conversation migrated to the topic of time and she said, "In your country time is money. In France time is love. In Italy time is style and beauty. But, in my country time is time!" She implied that her people attached no particular value to time—it was just time. Translating that into the business world, you shouldn't be surprised if people show up late for meetings, or if they don't hurry to complete negotiations, or sign agreements, or finish projects, or reach destinations. Time isn't important. It's just time. The differing cultural perspectives on time explain the divergent opinions when it comes to the urgency for approving projects, or granting permits, or making deliveries.

Similar disagreements can develop when trying to set goals. Many North American companies use Management by Objectives (Drucker, 1993) to focus employees' work on company results. It's part of their culture. In fact, Chapter 1 is based on a simplified Management by Objectives approach to get the project engineer started in his or her new assignment. North American companies tend to set definite targets and drive to achieve them by a specified time. Many are totally driven by short-term earnings. But in other cultures, goals may be more general—without a specific time frame. Perhaps more emphasis is placed on the process of getting to the end point than the goal itself. Having everyone from top to bottom in the organization aligned with each decision along the way may be more important to the participants than achieving the goal on time, or even at all. When such diverse opinions exist over something as fundamental as the underlying goals of a business venture, there is bound to be misunderstanding and, potentially, conflict. The home office back in New York is screaming for results while the people in the Far East are still trying to agree on the goals.

AN APPROACH TO RESOLVING DIFFERENCES

When working internationally, taking a long-term view and having patience is essential. It takes time to work through the process of understanding each other, resolving differences stemming from culture, and reaching agreement. Sometimes, the laws of the host country may even have to be changed for a deal to go through or a project to go forward. Whether working at the senior management's level or project engineer's level, there are three basic steps that can lead to successfully solving problems—or better yet, avoiding them:

- Communicating respect and setting the right climate
- Establishing a collaboration
- Creating a dialog, resolving differences, and reaching agreement

Respect and Setting the Right Climate

The eight cross-cultural communication skills enumerated by Lane, DiStefano, and Maznevski (1997) earlier are the way to get in the door. As you recall, right at the top of the list is being able to communicate respect. Respect, sound interpersonal skills, a sincere smile, plenty of conversations, and time are vital to creating a positive climate for discussions. In this type of constructive environment, people feel encouraged to listen to others' views. New ideas sprout and grow. Differences shrink and common thoughts multiply. If the other party is dealing in good faith, there is a better-than-even chance you will be able to find some common ground to work from. Whether you are negotiating a deal, planning a project, forming a joint organization, or solving a problem, a positive climate promotes success.

Establishing a Collaboration

Establishing a way to collaborate is finding the *structure* in which to work. It's not the actual dialog itself—that comes later. Let's consider an example to illustrate. You and your host country counterpart (we'll call him TC) are tasked with building a site to assemble custom-built equipment skids that will be sold to the construction industry in Southeast Asia. Components will be sourced locally and internationally.

You stop by TC's office to discuss the new project. Instead of jumping right into the business, you decide to exchange pleasantries first. In his language, you say, "Good morning. How are you?" That's always a good way to show that you are reaching out and trying to fit into his culture. Learning and using the local language will start you on the road to gaining rapport.

Then you take the opportunity to ask questions about the upcoming national holiday celebration and how his family usually celebrates it. TC sees your interest and describes the activities from dawn until dusk. Eventually when your conversation is winding down, you gradually steer it to the project. You sense that he is enthusiastic about it but doesn't know where to start.

After a long pause in the conversation, you suggest forming a task force consisting of you and him as coleaders and several key people from each of your teams. After thinking about this for a moment, TC suggests that maybe there should be a Big Team consisting of you, him, and your respective bosses in addition to the Small Team, which you had just described. You pick up on his desire to have his boss directly involved in the decision making and agree with TC's suggestion.

To keep the momentum going, you volunteer to write a simple charter to explain to your bosses what you propose and ask for their approval. You pick up a pad of paper and offer to write the proposal containing the following points:

- Objectives
- How the task force works
- Who will be on the Big and Small Teams
- When the task force's recommended plan will be ready

TC studies the four points and expresses concern about the word "objectives," so you change it to "what the task force will do." He smiles and you think you have his agreement.

At this point you propose a kickoff meeting, and TC agrees. He says he will arrange a dinner at a nice local restaurant but wants it to be after the holidays. He claims he needs the extra time to propose this to his boss. You know your boss won't like the delay, but you reluctantly agree.

You're now 90% of the way toward creating the structure in which TC and you can collaborate. If you both continue to maintain a positive climate and work as equals, TC and you will *own* the results and be committed to implement the solution that the task force recommends. The link to upper management that TC suggested will keep the task force's results on track and will help ensure that the recommendations will be compatible with the overall business venture.

In hierarchical societies, the link to management is crucial since no one will do anything until the boss says so.

Of course, this example is just one way to collaborate. There may be other structures, such as workshops, seminars, conferences, and committees. The point is to constitute some type of joint effort and give it direction. That way the interaction takes on a life of its own.

But as they say, the devil is in the details.

Creating a Dialog, Resolving Differences, and Reaching Agreement

With the climate and structure established, it's time to talk. We'll call the *process* of exchanging and discussing ideas the *dialog*. In practice, it's virtually impossible to separate the dialog from the processes of *resolving differences* and *reaching agreement*. As you work through your agenda, disagreements and misunderstandings are bound to occur, and they have to be recognized and addressed. The three are intertwined and progression toward agreement comes in increments.

Since these are processes, *it's just as important to focus on how the discussions are handled as on the subject matter*. The eight cross-cultural communication skills (respect, accepting that your own views aren't universally regarded as valid, being nonjudgmental, tolerating ambiguity, expressing empathy, being flexible, taking turns, and admitting what you don't know) are called into play as you seek to understand what is spoken and the intentions behind the words.

Let's illustrate this by continuing our example. It's now been 3 weeks since you gave the draft of the charter to TC. You've called several times, but he hasn't returned your call. The national holiday has come and gone, but there has been no response. Since the kickoff meeting is tomorrow, you decide to go to his office unannounced and try to get agreement on the charter.

TC is glad to see you. Even though you're irritated, you ask about his holiday celebration. His face lights up with a smile as he tells about his brother, whom he hadn't seen in years, coming for a visit.

You gradually bridge the conversation to business and ask, "What does your boss think about the charter?"

TC looks away and doesn't answer. He probably hasn't even shown the charter to his boss, but you decide not to corner him on that issue.

You sit down at his office table, and, as he sits down across the table, you take out a copy of the charter.

"We have the kickoff meeting tomorrow and I was wondering if we are in agreement on the charter?" you ask.

There is a long silence and you resist the urge to fill it.

Finally TC smiles and says, "Objectives are very difficult."

You look at what you had written:

What the task force will do:
- Develop functional requirements by end July
- List of facilities, buildings, and equipment by end August
- Develop plant layout by end September
- Submit recommendation by mid-October

After asking a series of friendly questions, it's pretty clear to you that not only do you not agree on the objectives, but *TC doesn't want to say no to you, directly.* That second thought puzzles you, but you store it for future reference and press on with trying to solve the problem with the objectives. (In TC's culture, the concept of "face" is highly valued. Saying no to you will cause you to lose face. Since he respects you, he doesn't want you to lose face, so he says nothing rather than saying no.)

After a long silent pause TC says, "My boss doesn't need to know about all that detail."

You think about this. *Could this mean that he doesn't want to share his specific objectives with his boss?* You'll probably never know.

After more discussion you take another tack. "What if we had these objectives?" You line through the objectives, write two bullet points in the margin, and then pass the charter to TC:

What the task force will do:
- Have a good idea of what the plant will look like
- Meet with the Big Team to discuss it

You know that you can influence the work once the task force is underway, and with some effective persuasion you can probably accomplish your original objectives.

"I like those objectives better," TC says.

"What about the rest of the charter?" you ask.

"It looks fine but it may take a little longer," TC says.

"What if we add another month to the schedule?" you ask.

"That might help," TC says.

You make the change on the draft. You then ask, "Will you have to review this with your boss or can we hold the kickoff meeting tomorrow?"

"Let's hold the meeting," TC says.

After you leave his office you're relieved that you have cleared the roadblock, even if it did take 3 weeks and you had to lower your standards on the objectives and schedule. You scratch your head and wonder how TC can see things so differently. It's little consolation that he is probably wondering the same thing. It would be helpful to know what his culture is and where it differs from yours.

Some Insights into Cultures

Understanding cultural differences and how they impact international business is a complex social science, and a thorough coverage is certainly beyond the scope of this book. Nevertheless, some insight into the basic variations in the value systems of human cultures can point out where to look for differences and trouble. Lane, DiStefano, and Maznevski (2006) explain six cultural orientations proposed originally by Kluckhohn and Strodtbeck (1961):

1. Relation to the environment
2. Relationships among people
3. Mode of human activity
4. Belief about basic human nature
5. Orientation to time
6. Use of space

These cultural orientations are useful to a project engineer since they reflect how cultures solve problems, address issues, and even approach business. At first appearance, the list seems somewhat obtuse, but Table 8.1 concisely summarizes the essence and relevance of this framework's meaning. I recommend that you refer to the book (Lane, DiStefano, and Maznevski, 2006) for a rather extensive description of those cultural orientations if you're interested in pursuing them further. The following sections will merely acquaint you with their subject matter.

1. Relation to the Environment

Some cultures see themselves as subjugated to the forces of nature—*what will be, will be, and there is no way to influence the outcome.* Their goals tend to be vague and qualified with conditions. At the other end of the spectrum, there are cultures that feel they have mastery over nature. Their goals are more specific and confident. Somewhere near the middle of the scale are cultures that live in harmony with their environment and have a high regard for it. Their goals would reflect that value.

2. Relationships among People

One of the most basic and complex values in society is how the people relate to each other. Their common view can be predominately individual based, group based, or hierarchical (status based). More specifically, these views involve how much responsibility a given person feels towards others.

The early immigrants to North America typify the rugged-individualist attitude of the individual-based view. They were independent, self-reliant people who

Table 8.1

Summarized Cultural Orientations Framework (Lane, DiStefano, and Maznevski, 2006)

Issue	Variations		
Relation to the environment (to the complete environment and to life and work in general)	**Subjugation** • Predetermined • Inevitable • Accepting • External control • Dependent • Fate	**Harmony** • Interdependent • Coexist • Live together with	**Mastery** • Control events/ situation • Make happen • Independent • Internal control
Relationships among people (power and responsibility)	**Hierarchical** • Vertical differentiation • Authority • Status (age, family, seniority, etc.)	**Group** • Horizontal differentiation • In-group, out-group distinction	**Individual** • Informal • Variable • Status (personal achievement) • Egalitarianism
Mode of human activity (mode of action)	**Being** • Spontaneous • Act on feelings	**Thinking** • Think and feel • Work and self • Seeking, becoming • Control self	**Doing** • Achieving/striving • Compulsive • Performance • Work is central focus
Basic human nature (changeable? unchangeable?)	**Bad** • Suspicious • Close supervision • Theory X	**Mixed/neutral** • Product of social environment • Consultation	**Good** • Information • Participation • Collaboration • Theory Y
Time—general orientation	**Past** • Respect tradition and proven ways • Precedence • Maintain continuity	**Present** • Current realities • Near term • Respond to change	**Future** • Longer term • Anticipate change
Time—activities	**Polychronic** • Relaxed • Elastic • Less critical		**Monochronic** • Tight schedules • Punctual • Hectic
Use of space (personal ownership)	**Private** • Closed • Secretive • Distant	**Mixed**	**Public** • Open • Physically close

took care of their close family members. When this value is dominant in modern societies, organizational structures tend to be somewhat informal and flexible. People tend to communicate with whomever they must across the organization in order to get the job done. Teamwork tends to be voluntary and informal.

The group-based view is common in some Asian and Mediterranean societies. Individuals value being a member of a group or tribe and feel responsibility for other members of the group. Teamwork within the group is natural and routine. Communication is mainly confined to members of the group, and rewards for achievement would be based on group rather than individual accomplishments.

In status-based cultures, such as aristocratic societies or caste systems, individuals identify with groups that are nested in a hierarchy. People feel responsible for those within their group, and there is both vertical and horizontal differentiation between the groups. In the business setting, the organizational structure is probably pyramidal. Communication and influence tend to follow lines of authority, and teamwork is more formal and structured than in societies with group-based or individual-based relationship value orientations. Rewards would tend to be based on status within the organization.

3. Mode of Human Activity

The activity orientation affects how people of a given culture view their work. On the *doing* extreme, people have a strong work ethic—they are the workaholics. Societies in the middle take a more reasoned approach and are able to balance work and leisure. On the *being* extreme, workers tend to be more spontaneous and operate from their feelings, rather than a compulsion to achieve. This orientation generally affects decision criteria, reward systems, concern for output, and the types of information and measurement systems that are implemented in the workplace.

4. Belief about Basic Human Nature

There are two aspects to this value orientation. The first (column one of row four, Table 8.1, under the title of the row) considers whether the societies think human beings are changeable or not. The other (columns two, three, and four) considers whether the basic nature of humans is good, mixed, or bad. In the middle of the scale, societies view people as neither good nor evil, or a mixture of both. Those value orientations tend to show up in the management style and amount of control organizations exert over their employees. If the organization views employees as basically good, it will give them more influence in performing their job and making decisions (Theory Y management). If the organization sees the workers as lazy and unwilling to perform, they will resort to a more authoritarian form of management (Theory X management).

5. Orientation to Time

This concept occupies two rows in Table 8.1. The first depicts the three aspects of the general orientation to time. Cultures with a time orientation that stems from the past depend heavily on precedence and history when solving problems. Other cultures tend to focus on the present and the immediate effects of the issue that they're trying to solve. Still others see the long-term consequences of the situation as important and weigh them heavily in their decision-making process. Cultures' orientation toward time affects their planning horizon, with the future-oriented cultures having the longest.

The other time orientation has to do with how the people approach their activities. In this orientation there are only two dimensions. The punctuality of the Swiss or German train systems typifies the monochronic orientation. The other extreme, polychronic, was indicated by my tour guide when she said, "Time is time."

6. Use of Space

The final cultural issue relates to how people regard the ownership of the space around them. In the workplace, those with a private orientation would tend to prefer individual offices and perhaps a closed door. They would also prefer one-on-one interactions and communication to get their work accomplished. People with a public orientation would prefer an open-concept office plan where they can frequently interact and communicate with several others at the same time. In the middle of the spectrum are those with a mixed attitude that prefer some privacy but also like areas in the office where they can meet and talk to others. Their communication would be more constrained than people with the public disposition. With the variety of cultural diversity in our workplace today, the orientation to space can cause serious workforce issues if it is disregarded when laying out the office floor plan.

Using the Cultural Orientation Framework

Lane, DiStefano, and Maznevski (2006) suggest using Table 8.1 to help map where differences might occur. Using two different colors, you can mark the dominant characteristics of your and the other person's cultures to see where the similarities and differences lie. Of course they are quick to point out that in any society there is a diversity of values and views, so this is just an indicator. There is no substitute for getting to know the people you're dealing with.

One final caution about Table 8.1: Don't try to line up the columns and make any descriptive sense of it. The table's vertical alignments aren't meant to illustrate variations in value orientations.

With this bit of fascinating and eye-opening theory behind us, let's consider how to apply all of this to the project engineer's job in a foreign setting.

APPLICATION TO THE PROJECT ENGINEER'S JOB

GAINING RAPPORT

During the war in Vietnam, I was assigned as an adviser to a Vietnamese engineering battalion. Upon arrival up-country, my commanding officer told me that my first priority was to *gain rapport* with my counterpart—the Vietnamese officer who commanded the battalion. Since then, when I've been in international situations, I've often reflected on the wisdom of that advice. Rapport is a relation characterized by harmony and accord, which is difficult to achieve under the best of conditions—let alone in an intercultural setting with its potential for misunderstandings. But the idea of *gaining* rapport is achievable and certainly worth striving for. It's an extension of what we discussed earlier:

- Getting to know the people you work closely with
- Taking an interest in them
- Listening to their ideas
- Working together on the job—suffering the same difficulties, disappointments, and dangers

It seems that if you and your counterparts have developed a mutual respect, you have gained rapport.

On the other side of this same topic, probably the quickest way to destroy the rapport you've gained is to say or even imply that you know more than your counterparts do. No one likes to be told that you are smarter than they are—especially if they think it may not be true. For example, you may not have as much information available as they do, and your point could be naïve, irrelevant, or completely countercultural in their world.

In the early stages of the job, the people side is far more important than the task side in the international setting. Rather than demanding that a certain course of action be taken, suggest a solution. Listen to your counterparts' ideas and try to reach a consensus. Plant a seed, water it, and let it grow. Take it slowly and deal with the easy problems first. Then, when you sense that you and your counterparts are attaining a degree of rapport, you can move to collaboration and solving the more difficult issues. Consider gaining rapport as an essential part of creating the right climate for the dialog that will take place when planning your project with your counterparts.

INTERNATIONAL PROJECT PLANNING

At any level, from top management down to the project engineers, the most natural structure for collaboration on international projects is the planning process. Both your organization and the host country organization have project agendas to achieve. There is no better way to reconcile those agendas than a formalized, joint planning exercise. It's good business to get key host country leadership and operations personnel fully engaged as early as you can. Safety, operability, local content, construction, engineering, purchasing, and regulatory approvals all run smoother if local considerations are taken into account and local people are involved.

Safety Planning

One of the critical challenges for U.S. companies on international projects is safety standards. During planning, two factors are crucial to achieving a safe work environment:

- Management commitment
- Planning to have employee involvement in the safety program

In our earlier example, TC and you are running the small project—you are the management. Early in the planning process, the two of you can jointly publish a short one-page bilingual safety philosophy that outlines your mutual commitment to preventing all accidents in the workplace. Hang it on a wall of the conference rooms and other places around the offices as a constant reminder. Place other safety slogans and warnings visibly around the workplace.

As people begin to mobilize on the project, TC and you can bring a safety professional onto your team to help write safety procedures and indoctrinate the workforce. The program should include local knowledge, international standards, and employee participation to help the team members analyze and improve the safety of their own jobs. Once the project starts, TC and you should walk through the job site periodically (monthly or quarterly), talking to employees and supervisors about safe working practices and making the necessary on-the-spot corrections. You can both follow up on your findings at your regular staff meetings.

Operational Philosophies and Requirements

The supervisors who will run the plant or facility should have input to its design. They are the ultimate customers of the design and construction process. But design engineers, if left to their own devices, will tend to repeat previous designs, which may not be best for a facility in this host country. Take, for example, the control system for a chemical plant in a developing country. Should the plant be highly automated so that relatively unskilled workers won't have to intervene in the process, or should it be low tech so that operators can be more easily trained to run the plant? Can a sophisticated control system be maintained in the host country, and are parts and service available? Only after discussions with people who know how to run a plant in the host country can these questions be answered. There are also important nontechnical issues that personnel with local knowledge must address. Plant leaders should have a say on considerations, ranging from the flow of materials through the plant when it is operating, to lunch facilities and break areas for multicultural workforces. These are typical of the operational aspects that must be factored into the design.

Getting the operational personnel and the designers together will head off problems down the road and facilitate the process of getting cultural differences out on the table and resolved. Organized reviews of the operating philosophies and an operational design review at an appropriate point in the design will promote sound operational input. But don't be surprised if the operational people can't visualize the plant from a set of engineering drawings. Extra measures such as scaled models of the facility or 3D CAD animations may be needed to show the operators what the plant will look like. These discussions will take time and patience but reward you with the valuable input that you need.

Local Content

Persons from the United States without international experience are often blind to the *importance of doing work within the host country* on international projects. The United States is probably the only major nation in the world that operates a free-market economy for foreign ventures within its borders. Other countries require projects within their jurisdiction to do some or all of the work through local companies. This creates jobs and builds their economy. Officials sometimes mandate a certain amount of *local content* in terms of the value spent in-country, work-hours used locally, the training to be provided to local workers, and sometimes the transfer of technology to the host country.

During the planning process it's wise to proactively optimize (not maximize) what can be done locally versus what must be sourced internationally. Using local labor, particularly in developing countries, may actually be good business as long the risks are managed and the necessary measures are taken to assure quality and productivity. And when it comes to running the plant or facility, the number of expatriates must be kept to a minimum to avoid the high expense. The cost to move and maintain an expatriate in a foreign location can easily be two to three times the cost of equivalent work in that person's home country.

Local content is a bona fide national interest of the host country, even though it is often carried to extremes. The best way to approach it is through a proactive prequalification of the local companies that your project can potentially use. This involves explaining the prospective work to the companies and requesting that they submit questionnaires with data about their experience, people, and capabilities to do the work. That way you will learn where their goods and services can be used, if at all. Meanwhile the unsuccessful companies will learn in which areas they must improve.

On the fringes of local content, particularly in developing countries, you may find individual interests involved that can lead to corruption. Both government officials and employees of firms can be involved in corrupt practices, so keep the lessons of Chapter 7 in mind and make your judgments and decisions with integrity. Because of all these national and individual interests that are layered on top of the normal intercultural issues, communication isn't usually very clear. As a rule though, if you don't receive the necessary governmental approvals for your local content plans in a timely manner, you probably have the wrong answer and should make another proposal after gathering as much insight as you can.

Construction, Engineering, and Procurement Planning

Local construction methods have a strong influence on construction projects abroad. And—as you remember from Chapter 4—the problems in engineering and procurement create problems in construction. Thus, if the local influences aren't factored into engineering and procurement, their absence will have an adverse impact. Let me illustrate.

A contracting company was constructing an oil pipeline in a remote location. The work—including conventional, cylindrical oil-storage tanks—was designed by an international engineering firm. The tanks were designed to be constructed on site as they would normally be in the engineering firm's home country. When it became time to build the tanks, the contracting company hired a local firm to do the work. Shortly thereafter it was discovered that the local firm planned to prefabricate the tanks in large sections, then roll them up like a roll of paper towels, truck them to the site, unroll them, and weld the large tank sections together to form the cylindrical tank walls. This creative approach saved many hours of expensive welding at the remote construction site. Unfortunately the tanks had to be redesigned for the new construction approach, resulting in months of delay and disruption. Had a local firm been involved in the design, the problem would probably have been detected on the drawing boards at a much earlier stage, without creating a delay in construction.

Similar effects can occur in procurement. Take, for example, a project for which steel, wire, piping materials, and other bulk materials are purchased from international sources far from the construction site. In the latter stages of construction it's usually necessary to order additional materials to complete the job, due to rework or misestimation. Serious delays occur if the additional materials must be purchased and shipped halfway around the globe to meet an immediate project need. If the materials had been originally purchased locally or in the region, this type of delay could have been avoided.

Other factors affecting foreign construction are productivity and labor rates in the host country. These considerations influence the estimating of cost and schedule and execution of the job.

Developing countries have much lower all-in labor rates than industrialized countries and are usually far less productive. One result is that many more laborers must be mobilized to the construction site. Another is that the construction methods are usually far more labor intensive in developing countries. Let's define some terms and look at this more closely.

Think of the unit installed labor cost (cost per metric ton) for constructing a particular facility as the product of productivity and the all-in labor rate.

$$\text{Unit installed labor cost} = (\text{Productivity}) \times (\text{All-in labor rate})$$

where

- *Unit installed labor cost* is the total labor cost of the facility divided by the weight in metric tons,
- *Productivity* is the total number of work-hours directly applied to the facility divided by the weight in metric tons, and
- *All-in labor rate* is the labor cost of the facility divided by the number of direct work-hours; it includes the salaries of the various types of workers and the cost of equipment, supervision, training, and other costs directly involved in the work on this particular part of the facility.

In highly industrialized countries, workers are paid more and tend to use fewer hours to do the work. To be competitive on the world open market, they use sophisticated equipment and construction methods, which further increase the all-in labor rate. Today, a ballpark number for all-in labor rates in those countries is $45 to $55 per hour.

Developing countries typically pay construction workers $2 to $5 per hour at present. When the cost of direct construction equipment, training, and supervision are added, the all-in labor rate would be on the order of $15 to $20 per hour. However, productivity in direct work-hours per ton can be two to three times higher (less productive) because they tend to use more labor-intensive construction methods.

The point of this discussion is that local construction methods must be taken into account when planning and estimating the job. Those methods also affect the execution of engineering, procurement, and construction. There is no substitute for involving local people in this planning process. It's good business, no matter how interculturally difficult it may be.

Regulatory Permitting

Regulatory permitting must be planned for on any project—international or domestic. However, internationally it may be more difficult to understand which approvals are needed. *The planning process must be pursued far enough to discover any regulatory showstoppers.* For example, if a project has mobilized equipment and personnel to clear the construction site but doesn't have the necessary governmental permit, there will be a delay.

People who lack experience in permitting tend to procrastinate. They are afraid to list all the required permits for fear of missing one, so they push the task in front of them. Or they can't get a straight answer from the host government on what permits are required. One way to break this deadlock is to bring in an expert with local knowledge. Your counterpart should be able to help put you in touch with one.

The approach we discussed earlier in this chapter is an effective way to work even with regulators. Establish the right climate, initiate collaboration, and establish an ongoing dialog to resolve issues and problems.

However, in developing countries, there may not be a clear process in place. *This calls for a proactive dialog that gives the host country officials what they need to approve your project.* For example, the host country may not have a procedure for conducting an Environmental Impact Assessment. In that case you can bring in your company's expert (or a consultant, if necessary) to develop a fit-for-purpose, no-frills process that meets your requirements and accepted international standards. This proposal can be submitted to the local officials to get their input and hopefully their eventual approval.

This should give you an idea of how to deal with regulatory permitting on international projects. As with all of those planning considerations mentioned above, your International Tool Kit, plus patience and common sense, will help you succeed.

THE INTERNATIONAL TOOLBOX

Keep in mind that cultural interactions are highly complex. This chapter merely acquaints you with some basic issues present when dealing with people from other cultures that you may encounter over the course of your career. Even complete texts on the subject only reflect the perceptions and experience of their authors. While a detailed knowledge of the social science of intercultural dynamics would be useful and would add depth to your understanding, it is out of reach for most of us. *What's crucial is having a set of basic tools to deal with intercultural situations when we're thrust into them.* The key tools are the following:

- Awareness of and sensitivity to intercultural issues
- A commitment to understand where the other person is coming from
- A set of cross-cultural communication skills
- A basic approach that will help you deal with cultural differences:

- ○ Setting the right climate and gaining rapport with your counterparts
- ○ Establishing a framework to collaborate in
- ○ Understanding how to create a dialog, resolve differences, and reach agreement on the issues of the day

And always remember that everyone sees the world differently.

REFERENCES

Drucker, P., *Management: Tasks, Responsibilities, Practices* (Harper Business, New York, 1993), pp. 430–442.

Kluckhohn, F., and Strodtbeck, F., *Variations in Value Orientations* (Row, Peterson, New York, 1961).

Lane, H., DiStefano, J., and Maznevski, M., *International Management Behavior: Text, Readings, and Cases* (5th ed.) (Blackwell Publishing, Oxford, U.K., 2006), pp. 23–56, 187, 265.

Lane, H., DiStefano, J., and Maznevski, M., *International Management Behavior: From Policy to Practice* (3rd ed.) (Blackwell Publishing, Cambridge, MA, 1997), pp. 3–19.

Chapter 9

Advice from the Pros

Some time ago I read a "proverb" on the Internet that goes something like this:

Good judgment comes from experience.
On the other hand, much of our experience comes from bad judgment.

Author unknown

This chapter seeks to eliminate the bad judgment part of this cycle by giving you the experience and good judgment of some people who have "been there" before you. We can certainly learn from our mistakes. A friend and experienced construction manager once told me: "A lot of contractors ask me, 'How do you find so many mistakes?' My standard reply is, 'Because I've made most of them myself.'"

It's my contention, however, that it's better to learn in a situation where things are being done right—and there can be a number of right ways to do things. If you are learning in a situation where the project is struggling or failing, you will have learned what doesn't work—but you won't necessarily know what *does*.

Even though no two situations are alike, you'll always find it helpful to listen to people who are successful at projects and apply the lessons they've learned. Certainly you will have to be careful when adapting their experiences to your own job. But, by using their starting point, you'll be farther along the road to good judgment, and, hopefully your occasional bad judgment won't steer you into the ditch.

To help you get started, I asked several experienced project people the question, "What advice would you give to a new project engineer?" You'll note that each answer has a different emphasis.

The advice is organized by the type of people who have given it. For example, the first major heading, "Advice from Young Project Engineers," is obviously from people like you on their first project engineering assignment. The subheadings distinguish between the various contributors in that category. "How to Approach the Job" is by one person, and "Foreign Construction Work" by another. The pattern repeats itself under each major heading with subheadings denoting separate pieces of advice from different individuals. Direct quotations or paraphrases of personal communications or interviews are indented.

Hopefully, this collection of wisdom will be a useful addition to your tool kit. A friend told me of a Russian proverb that states:

Most people learn from their own experience, the wise learn from other's experience!
Author unknown

ADVICE FROM YOUNG PROJECT ENGINEERS

Let's start with some advice from two highly respected young project engineers who are just starting their careers. The first, "How to Approach the Job," is a direct quotation from a young woman serving as a project engineer at a fabrication site.

HOW TO APPROACH THE JOB

As a recent new-hire project engineer, I offer three pieces of advice:
- Use the time to learn.
- Come prepared.
- Stay organized.

First, just being on a project will result in learning experiences, so really use the time to learn. Try new things, meet new people, and go outside your comfort zone. While on the project you should also find a mentor, who can answer questions and provide guidance when you need it.

Second, come prepared. The contracting strategy employed on the project will dictate the level of involvement with the contract. On the project on which I'm currently working, the lump sum contract has required me to know every detail of the contract to ensure the contractor provides everything we require. The specifications in the contract also must be adhered to and require familiarity.

Last, being organized can be a great asset, especially if you do it right from the start. I have found it is best to keep as little paper as possible because I have had to move around on the project. Most all files can be kept electronically. Meetings can be a great way to get things accomplished, but they can also waste time. Come prepared to the meeting, have an agenda, and stick to the [time and subject] limits. A final recommendation for organization is to cut down on emails when possible. We sometimes forget that phone calls can often be a more efficient means of communication.

FOREIGN CONSTRUCTION WORK

While traveling in the Far East, I had the privilege of having dinner with a young man assigned as a client's project engineer on a mega-project. Here is a paraphrased summary of the advice he had taken the time to organize during his busy schedule:

After two years in the home office, I was transferred to a shipyard. I consider myself fortunate that my boss helped me define my job. He has given me responsibility for overseeing some of the engineering and construction work, as well as some important engineering interfaces. He spent the time to give me an in-depth appreciation for safety and for the commercial side of working the interfaces.

My main advice is to *keep an open mind* and *ask questions*. There is always more than one way to design something. Consider alternatives and see if they can achieve the same design objectives. *Talk to people at all levels* in the organization, especially those near the workface. The suggestions of workers and supervisors can often improve efficiency.

Some other things I have found useful are the following:

- *Get to know your counterparts on the contractor's team:* Get to know the people that you deal with on a day-to-day basis. Seek opportunities on the job and socially. Those relationships will help you work through the stressful times when they come.
- *Assign priorities:* Assess the priorities of your major tasks or responsibilities. Rank them as high, medium, or low, based on their criticality to project success. This will tell you where to spend your time and help you to be more efficient. You don't need to consistently spend 12 hours a day on the job.
- *Trust:* Until proven otherwise, have some faith that the contractor's (or vendor's, or client's) people know what they are doing.
- *Safety:* Our Area Manager once said, "If production is King of the Construction Site, then Safety is God." Make a personal connection with the safety of the workers; care for their safety. Make sure your concern is real, and recognize good safety performance.
- *Ethics:* Ethical questions will arise. Be true to yourself.
- *Attention to detail:* Pay attention to details, but pick out the details that have to be worked correctly. All details are important, but you can't work them all.
- *Training:* Make sure you are getting the right training early in your career.
- *Working overseas:* If possible, get a book on the business culture of the country in which you are working. Ask questions and seek the advice of people who know the culture. This is especially important in the Far East, where the rules of behavior are different from those of Western societies.

ADVICE FROM A SENIOR EXECUTIVE

Senior management (vice presidential level and above in large organizations) is an excellent source of insight that can give perspective to young engineers' careers. In this section, a senior executive comments on the general topic of how to achieve results that add value to a company.

Achieving Results

I believe there is a tendency to want to be part of the herd when one joins up—overall a laudable goal. However, in a competitive environment, others may attempt to slow the efforts of a hard charger. You must stay true to your own values. If it is your desire to work a bit extra or in other ways go the extra mile, then do so. Do not be intimidated by the lowest common denominator in the group.

The importance of experienced coworkers as mentors can't be overemphasized. Their advice can be more candid and helpful, particularly in translating the supervisor's views or company's approach.

Concerning negative office politics—just don't do it.

In written material, presentations, or work plans, think through and focus on what the business objectives really are. The reason I bring it up is that new employees have a propensity to focus on what they find interesting as opposed to what the company really cares about. This is really the gift of learning what your boss is caring about rather than what you individually might choose to focus on. Personally, I think it took me years to try to think that way.

Above all, learn to take responsibility. You have to be responsible for the good decisions and share credit, which will spread wonderfully. More important, have the fortitude to take responsibility for your part in what did not go well and be prepared with suggestions on the path forward.

ADVICE FROM PROJECT ENGINEERS, MANAGERS, AND EXECUTIVES

This section contains common sense and insights on a variety of topics that are relevant for project engineers. Once again, this material grew out of the question, "What advice would you give a new project engineer?" It's offered by project engineers, managers, and executives who have served on large international projects and in several different industries. The only common thread is that I know them and trust their judgment.

Project Proverbs

The first advice is a priceless collection of guidance for young project engineers that an experienced engineering and construction manager has accumulated from his own experience, his friends, and other sources over the years. He wrote this in chart form, which I have paraphrased into this subsection.

Within reason, plan for the worst.

You cannot plan for an earthquake every day, but you can anticipate problems. Equipment can be delivered late to the construction site. A model test may have to be repeated because of inconsistent results or new design conditions. Studies undertaken after the start of construction could have an unfavorable impact ... the list goes on. You must always think about contingency plans and "work-arounds."

Trust but verify.

Ronald Reagan's well-known motto applies to project engineering. Trust what people give you and act on it, but do your normal checking and due diligence to make sure that it is correct.

Don't rely on organization charts.

This basic premise, which is found in most theory of organizational behavior books, is a good way to understand what's happening around you. Instead of naïve reliance on organization charts, get busy and investigate:

- Who has the knowledge?
- Who gathers the information?
- Who solves the problems?
- Who makes the decisions?

These functions aren't generally performed by a single individual. When you discover the answers to those questions, you will be better equipped to influence that organization.

Venture outside your office.

Whether at an engineering office, a fabrication yard, or a construction site, you need to meet with people in their office, shop, building, facility, or other workplace. By going to the other person's workspace, you'll learn things you otherwise wouldn't have found out. In addition, if you are solving a problem such as the allocation of space, you shouldn't do it unless you have walked the site.

Don't let distractions drive your schedule.

It's hard to keep interruptions by your team members, impromptu visitors, email, and a host of other distractions from consuming all your time and dominating your daily schedule. You have to set the priorities. Set aside some time alone everyday to think.

Pick good role models; seek out a mentor.

Enough said!

The contract is a tool, not a club.

Know the contract and refer to it as necessary. It's hard to beat experienced contractors (subcontractors) at contractual gamesmanship. The more you can avoid using the contract, the better. As an example, you are more likely to get the contractor or subcontractor to follow a quality management procedure if you can show them it is good business, rather than saying it is a contractual requirement.

Don't hold a grudge.

The urge to get even is overwhelming. Some clients, contractors, subcontractors, or vendors will lie, take advantage of you, manipulate you, and play a host of other dirty tricks. To them it's a game. Despite all of that, you must continue to be a player and not seek revenge because you will need their help. Of course, your dealings must be tempered by how they treat you, but in a controlled manner.

If your success depends on someone accomplishing something, then that is an interface you must engage.

Most projects are complex, and the contracts are interdependent. One cannot succeed without others succeeding. Any time your success depends on the success of another contractual party (external interface), you must manage that interface.

Internal interfaces (with other parts of the project team) can often be more difficult than the external ones, which usually require written interface agreements, have a structured process for conflict resolution, and receive senior management attention from both parties. Internal interfaces often "sneak up and bite you."

Learn who to trust, and in what setting.

Deception is a regrettable but very real part of contractual gamesmanship. Project team members will have different levels of honesty, and you will have to learn whom you can trust and whom you can't. The level of honesty may also depend on the setting. Some people will be more open in private or at an off-site team building—but some won't.

Develop contacts in whom you have confidence.

Paraphrasing the metaphor "Blood is thicker than water"—"Friendship is thicker than organizational ties." To put this into practice, develop your own personal network of contacts and experts you can trust and use as a sounding board. That network takes time to build—in fact, it is a lifetime endeavor—but it is a great resource.

Be a good spy.

Knowledge is power. Find out what a contractor's or subcontractor's other clients are doing and what their problems are.

Everything you need to know about project execution, your mother taught you.

Much of project execution is common sense and integrity. Don't hesitate to challenge the experts or think outside the box. Young engineers often find flaws in project execution plans that senior planners and experienced project team members have overlooked.

An Intercultural Aspect of Contracting

An astute project engineer who has lived and worked in several cultural settings offers the following advice about contracts that cross cultural divides.

In the United States we view a contract as the conclusion of an agreement. In many cultures signing a contract is just the beginning of the negotiation. Don't be surprised if your counterparts begin haggling over matters that are written clearly in black and white.

Project Engineering for Manufacturing High-Tech Equipment

During lunch with a project executive of a high-tech equipment manufacturing company, I was able to capture his philosophy about manufacturing and a wealth of information on how project engineers and systems engineers function within his firm. This paraphrases his comments. I've added a few thoughts in parentheses to relate his comments to this book.

My key advice to young project engineers is, "Never underestimate how much effort is required to plan and implement a manufacturing project." Let me explain.

Our company's primary goals are to
- balance safety, quality, cost, and schedule,
- always take care of the customer, and
- make a profit.

Our type of equipment requires a significant level of design and interface with other contractors. The contracts are placed on an EPC basis and include significant design effort after contract award.

As we look at the lessons learned, the major risks to this type of contract are mainly up front before the contract is signed, and in the early 3 months of defining the projects and supplementing the customers' design bases. These risks can be controlled somewhat by contracting strategies, contractual terms, written qualifications concerning the client's job specifications, the inclusion of contingencies, allowances and risk money in the bid, and a period of reimbursable engineering to define the work and reach agreement with the client on what will be delivered to the project and how.

However, even with those measures in place, it is very possible to find out that the costs are overrunning. This is usually determined when the equipment is manufactured, and by then it's too late. The change management system can give an alarm warning that cost trouble is coming. If appropriate actions and countermeasures are taken, the adverse effects of a cost increase or schedule delay can be softened. On high-tech jobs that require development work, there is always the risk that engineering and testing will turn up problems. That risk needs to be accounted for in the bid either by contingency or the use of previously designed and field-proven equipment.

Our company has spent a lot of money and effort developing and documenting a management system that helps us execute projects. This ensures that all project personnel understand the company's requirements in managing a project. It's organized according to the phases of the (manufacturing) project and concentrates on defining the Design Basis and establishing a Project Execution Plan up front, since that's when most of the risk occurs. We have gates at the end of each phase that the project must successfully pass before continuing. Project deliverables for a particular gate are completed and reviewed. Design reviews are called at critical points in the engineering to keep the work of both systems and package engineering sewed together. Coordination between the systems engineers and lead package engineers plays a big part in the success of our project management system.

One of the issues with a manufacturing company is that the focus is generally on "widgets" while the EPC projects require more of a systems focus. To manage this, the systems engineers and the lead package engineers both report to the Technical Manager. It's essential that they interact with each other so that the product will function as intended.

The systems engineers are responsible for specific systems (e.g., the control system) that run throughout the product. They link the engineering work together, so that their systems work as a whole. For example, some of their duties include

- technical oversight of the overall system layout and design, including P&IDs (piping and instrument diagrams),
- preparation and ownership of the system design basis, and
- coordination with a host of other groups including interfacing contractors.

Lead package engineers are technically responsible for a physical part of the final product (e.g., the manifold package) including all its internal systems. Their main role is to make the transition from the contract to lines on the drawings and words in the specs. The lead package engineers have dual reporting responsibilities. They report to the Technical Manager for package engineering matters, and they report functionally to the responsible worksite managers during manufacturing and testing. The lead package engineers have a long list of duties to ensure that their packages meet the technical requirements of the contract, and that the design deliverables are provided according to the project schedule. (They are a true project engineer in every aspect except cost control.)

STRUCTURING AND ORGANIZING ENGINEERING AND PROCUREMENT ON MEGA-PROJECTS

This approach for structuring and organizing the engineering and procurement work on multibillion dollar projects is offered by a former project manager who is now the managing director of a firm. It provides a glimpse of two ways that large engineering and procurement jobs are organized and how the project engineers fit into each organization.

Work Breakdown Structure

The first step in managing a mega-project's execution is normally to split the work into an appropriate number of work packages. A work package is a discretely defined part of the facility or a task. The work breakdown structure (WBS) consisting of all the work packages is used as the basis for planning, budgeting, progress reporting, and cost control.

Organizational Options

The objective of the engineering and procurement organization is to produce all documents and drawings that enable the client to procure the required equipment and material, and later construct, install, and operate the facility. The WBS is normally reflected in the engineering and procurement project organization. An engineering and procurement organization may in principle be organized in two ways:

1. An organization structured according to the WBS and the work packages. Work package managers oversee a group of related work packages and report vertically to the project manager. The required engineering resources are allocated to the work packages and are managed by package responsible engineers or project engineers.

2. An organization structured in such a way that all engineering disciplines are gathered into an engineering unit, and all project engineers into a project engineering unit. In such an organization the project engineering unit consists of project engineers that are each responsible for "getting forward" the required deliverables for one or several work packages in agreement and cooperation with the engineering unit.

The advantage of organizational structure 1 is that the work package managers have direct control over the deliverables to be produced and the resources that produce them. The disadvantages are that you may not achieve an optimal utilization of the engineering resources, and you may not adequately address

- the general engineering activities that are common for all work packages, and
- system engineering activities for systems that run through several work packages.

For large engineering and procurement projects, the second organizational model is generally used. To recap and expand, the project manager controls the following main units:

- The engineering unit managed by an engineering manager, which includes all engineering disciplines. This unit produces all the documents and drawings.
- The project engineering unit managed by a project engineering manager and consisting of project engineers that are responsible for getting forward the right drawings and documents in one or more of the defined work packages.

- The procurement unit managed by the procurement manager responsible for all the commercial activities related to procurement of equipment and material.
- The project control group under a project control manager that handles planning, budgeting, progress reporting, cost control, contract administration, personnel administration, and the IT system.

KEY LESSONS LEARNED FROM A HANDFUL OF ENGINEERING AND PROCUREMENT MEGA-PROJECTS

This insightful review of the engineering and procurement lessons learned from a number of challenging projects is excellent reference material for any project engineer who wants to tackle mega-projects. It comes from a seasoned project manager with a number of successful projects under his belt.

Three Golden Rules
I've come up with Three Golden Rules:
- Develop a good relationship with the client (starting with an alignment process).
- Keep a close eye to the bottom line, i.e., the money!
- Above all else, develop an organization that does the job for you!

Organizational Development and Alignment
Those first and third points take hard work and long hours at the beginning, but they pay off in the long run! Organizational development involves all members of the project organization (both client and contractor) in a continual program throughout the project's lifetime. Team building is one part of it, but learning more about communication, conflict resolution, personal behavior, body language, creating a project success vision, and a code of conduct for this specific project are all parts of an organizational development and alignment program. It must be led by the project manager, inspired by a desire that it will benefit the project, and also enrich the project team members by adding to their personal learning.

Further, kicking off an alignment process with the active use of expectation analysis is another key aspect of organizational development, whether between client and contractor or internally within your own organization. Having a mutual dialog regarding expectations in all aspects of the work normally establishes an excellent basis for a continued, mutually beneficial alignment process. In addition, it facilitates further definition of the scope of work. In fact, expectation analysis with all the major stakeholders is an excellent tool to identify what is expected.

The organizational development program also allows the project manager to come in contact with most of the project staff. That initiates contacts and builds networks, which allow the project manager to feel the project's pulse through communication with individuals throughout the project's life. The information from those contacts and unofficial networks enables the manager to actively influence the project (through the line organization) before problems are visible on the progress charts or bottom line.

Other Lessons Learned

Here are some other key lessons learned from major projects.

Organization that does the job:

- On major projects, my preference is to split the engineering into two—one "producing" part (i.e., the engineering organization with all its disciplines and specialists) and one "internal client" part (the project engineering organization with project engineers as internal clients for their areas or modules). On engineering, procurement, and construction (EPC) projects, subsequent to the engineering phase, selected project engineers can move on to become an area or module construction manager, with full insight and influence from the engineering phase.
- The dualism in the engineering/project engineering organizational approach has the effect that the project engineers act as the project manager's prolonged arm, thereby providing the project manager with direct insight into engineering, without relying on the engineering manager's view of the state of progress and quality.
- Any project organization will work, as long as the key members of that organization agree that it is the right organization for the task at hand. That, however, is achieved only through full focus on team building and alignment—not only at the start of the project, but continuously through the project and especially when moving from one phase to the next.
- In most client-contractor communication there is a degree of give and take. If necessary, it can be useful to have a good guy/bad guy approach. However, the project manager must ensure that he or she *never* ends up as the bad guy. That role should be left to someone else on the team.

Planning that works: Planning must be done by the doers (lead engineers and supervisors), and not by the planning department alone!

Importance of front end loading (spending time, money, and effort on project definition):

- It is the depth and quality of project definition during the early phases that decides success or failure. By investing time and work-hours on project definition in the period prior to heavy project spending, the client will harvest dividends at the end of the job.
- Front end loading should be in every company's project execution model, along with constantly ensuring that engineering is fit for construction, commissioning, and operations. There must also be full focus on the completion and warranty aspect of the work at the end of the project and beyond.

- Using the most experienced EPC contractors in the definition phase will yield dividends throughout project implementation. Their substantial conceptual design and FEED experience coupled with a proven and documented way of executing projects is an asset. As a wise person said, "Good proven experience and know-how is priceless; inexperience costs even more!"
- Major subcontractors and vendors should be involved as early in the FEED as possible. They are the experts in their areas and will provide invaluable advice if handled in the right way.

Quality assurance:
- To me, the most effective engineering and procurement quality assurance consists of self-checking in every level of the organization.
- The other aspect of quality assurance is the initial Contract and Design Basis Review, which is a crucial part of starting any project. That effort aligns the technical, commercial, and contractual *expectations and objectives* of both the client and the contractor. To be more exact, it starts the alignment processes between client-contractor and contractor-subcontractor, which must be shepherded the whole way.

Resolving major internal organizational conflicts:
- On major projects, there will always be conflicts brewing, but don't make it your business to resolve them all. Some will resolve themselves, and others will be dealt with by members of your management team. The remainder is up to you to solve.
- When you must get involved, a good rule is to never wade in and resolve conflicts by fiat. If you do, you will become part of the conflict and will have to take sides. Make the parties come to a recommended solution (or separate recommendations if needed), which you can review and conclude in a timely manner, after possibly seeking advice from other project managers or executives.

PLANNING A HIGH-TECH, GLOBAL IT PROJECT: MANAGEMENT SUPPORT AND BUY-IN

Reengineering the business systems in a major, global corporation is a challenge that requires business judgment, political savvy, interpersonal skills, common sense, intelligence, luck, and a lot of perseverance. Here's how one executive approached the daunting task of getting senior management's support and buy-in for his software development project. Hidden in here are nuggets of truth for project engineers.

The Project
My project involves creating a database of customer information across our international business units, some of which were recently acquired. The project is

not a revenue generator, per se, but it adds value because of its enabling characteristics. This database becomes the basis for essentially anything that is done with any account. Each customer is entered in the database once and given a unique identifier. Data is continually updated across the globe while maintaining privacy, security, financial controls, business checks and balances, and compliance with local laws and regulations. There are a number of other boundary conditions, but probably the most crucial is that the database must serve the needs of our international offices and corporate stakeholders.

Working with Senior Management to Get Started

Fortunately, I had a senior management sponsor that underpinned the corporate commitment to the project. However, it took a lot of negotiating to get a strong team and an understanding that we would start small and phase the development of the database over time with decisions at each major milestone. To steer the project and create buy-in, we formed a Sponsors' Committee that consisted of stakeholders from across the organization.

The first phase consisted of a feasibility study to establish the possibilities and limitations. This phase identified a pilot project as a first step that would flush out the issues and bugs before going full scale. That way we could show the stakeholders where the project was going, step by step. They didn't have to make large commitments up front and could change course at the check points. We also kept the scope of the project, its benefits, and the project team's values in front of the management sponsors. Our project team rigidly controlled the cost and schedule for the feasibility phase and pilot phase to establish credibility with our sponsors. Our project reporting to the sponsors addressed two aspects:

1. Running the business
2. Changing the business

We had to demonstrate that 1 was getting better and 2 wasn't adversely affecting 1. Information also flowed the other way to keep the project team informed of management insights and issues.

Our success factors for working with senior management were the following:
- Start small and develop the project in phases.
- Develop the concept over time with all the stakeholders involved from day one.
- Maintain a deep and profound respect for the stakeholders, their views (even when they are different from yours), their problems, and what they stand for.
- Test ideas with the stakeholders many times. Use lobbying, diplomacy, and negotiation. Be persistent in a nice way.
- Show respect for the ideas of others. There can be no stupid questions, so answer each question with respect.

Those factors are probably applicable for many kinds of business initiatives where it's important to maintain senior management's support.

International Communication

When it comes to communication, whether international or within your own country, two phrases are useful:

- Seek to understand and then to be understood.
- Say what you mean and mean what you say.

"Seek to understand" partially involves figuring out what has to be done on the project. From a communications standpoint, it means understanding the agendas of all the stakeholders and what drives their business environment. Understanding the politics in the various organizations, the personalities of the people, and how they think and respond are all a part of understanding what *they* mean. It takes time to master this, but once the project manager understands what has to be done on the project and what the others mean, she or he can begin to effectively craft the message—"to be understood."

"Saying what you mean" takes on new dimensions in international companies. When the project manager understands this particular challenge, he or she can approach each group or individual in the most effective way, choosing the right words and arguments. This means influencing the stakeholders who will authorize the project, fund it, steer the work, and commit to implement the solution in their organizations. It also means being more effective at garnering the support of peers that will be called upon to provide resources or occasionally give their support to senior management. Equally important, the project manager will be able to effectively lead and motivate the project team that must define the business requirements, translate them into software, and implement it on the computing system.

"Meaning what you say" has all the normal connotations. It includes following through on the project's plans and commitments along with maintaining transparency and credibility.

One final note on communication with senior management: the project manager should anticipate the sponsors' questions and have the answers at her or his fingertips.

And, on the topic of seeking advice from senior management, my boss once said, "Don't come to me with a question without having an answer in mind."

Other Advice

On a more personal level, here are some techniques that I have found helpful. First of all, create time to think on a daily basis. Concentrate on a topic and let your thoughts flow. The topic can be anything from what to say at an important meeting to how to handle a problem. It can take as little as 15 to 20 minutes but should be free from interruptions. The place is unimportant: the office, the airport, on a train, lunch alone, a stroll with your dog. The results will be a lot of good ideas and a reduction in stress level.

Meetings, especially with upper management, are always important. When preparing for meetings, don't forget to find out who will be there and what their jobs are.

Finally, never stop learning. I've read more books in the workplace than I ever did in school.

QUALITY: AN HISTORICAL PERSPECTIVE

This quotation was related to a friend of mine by a senior executive, many years ago: "Years from now no one will remember how much it cost or how long it took. But they will remember if it didn't work."

BALANCING QUALITY, COST, AND SCHEDULE

This down-to-earth explanation of the balance between safety, quality, cost, and schedule brings this concept to life. It comes from a project manager with a diverse background in projects, drilling of oil and gas wells, equipment development, research, consulting, and expert legal opinion:

The objectives and outcomes (results) of a project may be described in terms of three parameters: cost, schedule, and quality. Cost and schedule are straightforward concepts, and these outcomes are readily measured and compared with objectives. Quality is a more amorphous term that includes a variety of subobjectives/outcomes, some of which are often poorly stated as objectives or measured as outcomes during and immediately following a project. Depending on the nature of the project and the objectives of the owner, some of the suboutcomes that may be lumped under quality are functionality, operability, reliability, aesthetics, safety (both during and after the project), and environmental protection.

At its inception, every project has a certain amount of uncertainty regarding the achievement of its cost, schedule, and quality objectives. The amount of uncertainty varies depending on the nature of the project. Typically, research, new technology, and very large construction projects have the most initial uncertainty. However, even a project to construct and install a birdhouse in one's backyard has some initial uncertainty (What size? Buy a kit or build from scratch? Will it rain?) The amount of uncertainty should decrease as the project progresses.

An important concept about uncertainty is that you can't make it go away. You can and should minimize it by thoroughly planning, designing, and engineering the project, but there will be some things you simply can't know with certainty at the outset. You can, however, decide which outcomes are most important and manage the project to minimize the risk associated with one or perhaps two of the three outcomes.

A useful analogy is a three-lobed water balloon, the lobes representing cost, schedule, and quality. The water in the balloon is the project's uncertainty. One may squeeze on one or two of the lobes to reduce the amount of water in them. The water expelled will, of course, expand the other lobe(s). If one tries to squeeze all three lobes, he or she will get wet! This analogy could be extended to the suboutcomes of quality, but I'll leave that to the reader.

Similarly, one may choose to tightly control one or two of the outcomes of a project, but, depending on the project's uncertainty, the other outcome(s) are necessarily put at some risk. The belief that all three may be tightly controlled in a risky project is a delusion.

In theory, it would be very desirable at the outset of a project for the owner and contractor to prioritize outcomes, identifying which are the most important. This rarely occurs, probably because of a fear that the lower-priority outcomes will be allowed to run amok (go out of control). However, even a moderately perceptive project engineer or manager can soon deduce which outcomes are most important.

Often, priorities will change during the course of a project. If, for example, a project that was not originally schedule driven slips too much, schedule may become the highest priority until it gets back on track.

Increasingly, larger contractors are developing or adopting comprehensive project management systems. These systems have evolved over time and tend to treat all outcomes with the same degree of importance. This is understandable since these generic management systems must apply to a variety of projects with different priorities. The trick in employing these systems is to tailor them to the specific project at hand. If all aspects of these systems are applied with equal emphasis, the balloon will burst!

A new project engineer will not be in a position to control or significantly influence the project priorities or the amount of emphasis placed on implementing aspects of a project management system. The engineer should, however,

- realize that priorities exist whether acknowledged or not,
- identify what they are, and
- try to focus her or his efforts on the things that really matter.

RISK MANAGEMENT AND DEALING WITH CRISES

I once asked a construction manager, "What do you do when a project goes into the ditch?" He replied, "The right question is, 'What do you do when a project starts to veer into the ditch?' Maybe an even better question is, 'How do you avoid serious problems?' The answer to that last question is risk management."

RISK MANAGEMENT: A PROJECT MANAGER'S VIEW

This advice is from an excellent project manager who believes in risk management and practices it on all aspects of his jobs. It's the primary point he chose to emphasize for new project engineers.

You *must* manage the risks. Always have at least two ways to succeed or mitigate the risks. The steps are the following:

1. Identify the risks:
   - Safety
   - Commercial
   - Technical
   - Project execution
2. Identify the mitigation steps for each risk.
3. Develop a plan of attack to mitigate each risk.

RISK MANAGEMENT: AN ENGINEER'S VIEW

An engineer skilled in the design of high-pressure natural gas processing offers the following perspective on risk management.

Companies must manage risk, and the project engineer is on the front lines. The tools of the trade are

- understanding the duties for the facility (e.g., functional requirements, temperature, pressure, environment, the corrosive chemicals both inside and outside of the equipment, the effect of probable hazards),
- choice of materials and making sure they are manufactured, welded, installed properly, and used in the right place (the advice of expert materials engineering experts is often required),
- codes, regulations, and industry standards and their enforcement,
- company specifications and their enforcement,
- proper engineering, design reviews, construction practices, and verification,
- risk assessments to find and correct potential problems, and
- quality management.

The project engineer is the guardian of the requirements that are embodied in those tools and applied during engineering, procurement, and construction. The crucial importance of skillfully employing those tools is underscored when we read in the newspaper that something like using a pipe made from the wrong material for the design conditions has caused loss of life and extensive damage to property and a company's reputation.

RISK MANAGEMENT: BEGINNINGS AND ENDINGS

A seasoned project manager with many successful projects to his credit points out a key risk factor for project engineers to recognize and avoid.

Bad projects are often caused by bad beginnings and bad endings.

Poor technical or project definition by the client, ill-defined project requirements, a hastily prepared proposal with an inadequate risk assessment, a weak or incomplete project team, and poor project execution planning are all examples of things that can cause bad beginnings. Typically, bad endings result when incomplete work from previous phases is carried over into the systems completions phase (where it is more expensive to do), systems aren't completed in the right order, serious problems must be rectified (e.g., systematic welding defects), or the project team doesn't focus on completing the job.

These problems are far easier to avoid than to fix!

RISK MANAGEMENT: DEALING WITH CRISES AND CALLING TIME-OUT

All of the other information in this chapter is written by others or derived from personal communication with them. I make this one exception to round out the advice on risk management. Here is my best advice if you ever find yourself facing an impending project crisis.

When real problems begin to surface on a project, it is wise to call time-out, but this is easier said than done. Often, time-out has to be called from a level or two in the organization above where the problem is occurring. Sometimes there is a tendency for those closest to the problem to deny that it exists. They continue to incrementally try to progress the work by applying one corrective measure after another or simply working harder.

Once the seriousness of such a situation is recognized, an organized assessment is needed. Usually an assessment team from outside the project is brought in to objectively appraise the problem and recommend solutions. Depending on the nature or seriousness, a parallel audit investigation is instigated to determine if financial or business conduct infractions have occurred. Management then decides what changes will be made to bring the project to a conclusion. In extreme cases, a rescue team is formed to take over the troubled work.

RISK MITIGATION THROUGH ORGANIZATIONAL DEVELOPMENT AND ALIGNMENT PROGRAMS

One project manager in tune with the people side of projects has this to say about the importance of aligning all parties, including the client, to mitigate risks and avoid failed projects:

Today some of the focus has turned away from organizational and personal development programs (including alignment processes), possibly due to the ever-present focus on reducing costs. Regrettably, some of the project disasters we have seen in recent years can be attributed to just that. Development and alignment programs now seem to be receiving increased interest again.

LOOKING BACK: HOW TO APPROACH THE JOB

A senior manager in a large fabrication yard offers this advice to newly hired engineers:

> Many engineers directly out of school are a little unsure of themselves in their first job. Confidence is the key. Ask lots of questions. You are there to learn. Your mentor can help.
>
> Don't sit back and let others do the work. Participate. You may have more to offer than you think.
>
> Get along with those you work with. Be kind and helpful, and you will be more likely to receive the same in return. In addition, one of them could be your boss some day, so don't burn your bridges.

LOOKING BACK: THE PROJECT ENGINEER'S ROLE

A project engineer with over 30 years of international experience puts his role in perspective:

> Project engineering is the toughest job on the project. The project engineer doesn't usually have the "stroke" to force others to do what he or she wants done. You must use whatever means are available to persuade and influence them to help in accomplishing your goals.

LOOKING BACK: A CAREER STRATEGY

Here is some advice from a construction manager on career strategy for new engineers who want to broaden their careers and potentially move into construction and project management:

> I think it's best to start your career in engineering positions. That way you are grounded in the technical aspects before moving into construction.

This same advice could also apply to engineers in the manufacturing sector. A good way to make the transition from engineering to construction or manufacturing is to work on an area as a project engineer during the design phase and then manage the construction or manufacturing for that same area later on the same job.

AN INTERVIEW WITH AN EXPERIENCED PROJECT ENGINEER AND MANAGER

I'll wrap up "Advice from the Pros" with an interview that I think you will identify with. This person is now managing large construction projects and has the uncanny knack of visualizing the processes and forces at work on projects. He learned volumes from his grassroots experience as a young project engineer.

Q: Would you describe your favorite project engineering experience?

A: It was the engineering and construction of a large offshore oil and gas plat-form. I started as an engineer on the design team in Houston, then moved to the Far East as a project engineer on the construction team. Following that, I was the project engineer for the living quarters, where the offshore workers would eventually live, eat, and sleep. The living quarters job included the last phase of engineering, construction, outfitting, commissioning, cost and sched-ule control, accounting, and even regulatory approval. During engineering and construction, I had to cooperate with the operational people who would even-tually run the platform.

Q: Did you have a mentor? If so, did the mentor help or hinder you?

A: As a young designer, I had a mentor. He was always approachable and help-ful. Mentoring has to be taken seriously. In my current organization we evaluate senior people on that point. If a mentor doesn't do a good job, it hinders the new employee's development and may discourage that person from going to others for help.

Q: What were the things that made your project engineering job successful?

A: Aggressive follow-up:
  • Check as much as possible and be proactive.
  • Stay a few steps ahead (think about activities that are coming up).
  • Communicate with everybody—I learned a lot that way.
  By checking and verifying the engineering, I discovered a structural problem that could have seriously damaged the living quarters during lifting operations.

Q: What parts of the project engineer's role do you consider to be most important?

A: The project engineer has to know his or her boundaries. Asking lots of questions to get clarity is much better than trying to figure it out later, which is a common mistake.
  Safety, quality, cost, and schedule are crucial. Safety is a must. Those who think they can have all of the other three will most likely be surprised.
  Change control is also a major part of the role.
  And, of course, communicate, communicate, communicate!

Q: What essential skills should a project engineer have?

A: In addition to what we have already discussed, leadership is important. Planning and organizing are leadership, the rest is management. Planning is finding the best way to do something and the activities that are needed. Scheduling is linking the logic between the activities. I usually get others to do the scheduling.

   Leading is doing the right things. Managing is doing things right.

Q: How do you kick the job off?

A: Start by writing down the objectives for the team. I give them my vision by visualizing a successful outcome. Then I discuss the guiding principles of how I want the team to behave. For example: be part of the team, be honest, communicate, and tell the truth. I generally repeat those guiding principles at each major transition.

Q: What tools do you use to control the job?

A: I like to use leading indicators that predict future problems rather than lagging indicators that show what has happened. These are useful for safety, quality, and project execution.

Q: What are the major pitfalls that can get the project off track? What do you do to keep them from happening?

A: Lack of communication is a major pitfall. When teams are working within their own group and not taking time to interface with other groups, it's only a matter of time until problems show up.

   Another pitfall is a lack of alignment on objectives. This goes back to our earlier discussion about balancing quality, cost, and schedule. The project can't have it all.

   Another pitfall is really a caution to beware of false deadlines. The date that a platform leaves the fabrication yard is a real deadline. The start-up of a facility is a real deadline. Let me give you an example of a false deadline. I was on a team that was preparing a bid package. A date had been set to send the bid package out, but the package was incomplete when that date arrived. I insisted that we wait an additional 2 weeks until some important parts of the package were completed. This didn't make people happy, but we waited the 2 weeks before sending the completed package to the bidders. The quality was there, and the contract (the work that was contracted for) was delivered 1 month early.

Q: What do you do when things do get off track?

A: Call time out... sooner rather than later. Sometimes that's hard to do, but it has to be done.

Q: So far we have talked about tasks. Can you say anything about the people side?

A: Communicate!

Q: What tips or advice would you give a young engineer?

A: Of course, the young engineer must start by being ethical. There are some good books around on that subject. In addition, the new engineer needs to become aware of those aspects of business conduct that can create trouble.

   On the positive side, to be successful today, a person should do the following:

- Be technically astute and competent. That is the primary thing that gets her or him recognized.
- Be able to get along with people.
- Work toward the desired response, but don't remain silent. You have a responsibility to point out problems and risks. However, when a decision is made it is best to fall in line, unless it's illegal or counter to business ethics, financial controls, or company policies.
- Have a proactive attitude that you can get things done . . . a can-do attitude.

This is the voice of experience!

Some of you may be wondering about that last line. How do you have a *can-do* attitude when you're so new you don't even know what *to* do? That will be the subject of the next chapter.

Chapter 10

Approach the Job
with Confidence

*Whatever you can do—or dream you can do—*begin it!
Boldness has genius, power, and magic in it.

Attributed to Goethe

One of your primary assets in life is a feeling of confidence. It affects the way you think about yourself and the possibilities that confront you. If you are confident and expect a positive outcome, your intentions will bring energy and add reality to successfully achieving that outcome. If you expect the client to accept your change order request, he will be more likely to do just that. If you expect your boss to approve your recommendation, she probably will. We are not talking about a cocky brashness or wishful thinking, but a solid, calm mind-set that *you can do this*.

Your confidence will also empower the people around you, especially those you lead. If you expect them to succeed, they're more likely to succeed. That doesn't mean that your expectations are a substitute for their competence and effort, but it does help others see new options and strive to accomplish them. As mentioned earlier, trust is the grease that improves their performance and builds confidence.

In short, approaching the job with a confident attitude is a primary factor in being successful.

THE PYGMALION EFFECT

There is a body of scientific evidence which conclusively demonstrates that our expectations toward others are, in effect, a self-fulfilling prophecy. In summary:

- We form certain expectations of people and events.
- We communicate those expectations with various cues.
- People tend to respond to those cues by adjusting their behavior to match them.
- The result is that the original expectation becomes true (Accel-Team.com, 2005).

The Pygmalion Effect is named for a mythical Greek sculptor who fashioned a beautiful female statue that came to life, whereupon he fell in love with her. In the early 1900s, Bernard Shaw wrote the play *Pygmalion,* which eventually was adapted to the musical *My Fair Lady.* Phonetics Professor Henry Higgins wagers with his friend Pickering that he can pass off the Cockney flower girl, Eliza Doolittle, as a duchess. Eliza's comment to Pickering poignantly summarizes this powerful principle:

> You see, really and truly, apart from the things anyone can pick up (the dressing and the proper way of speaking and so on), the difference between a lady and a flower girl is not how she behaves, but how she's treated. I shall always be a flower girl to Professor Higgins, because he always treats me as a flower girl, and always will, but I know I can be a lady to you because you always treat me as a lady, and always will. (Shaw, 1916)

There are a number of interesting corollaries to the Pygmalion Effect, but one stands out as particularly relevant to this discussion: "The best managers have confidence in themselves and in their ability to hire, develop and motivate people; largely because of the self-confidence, they communicate high expectations to others" (Accel-Team.com, 2005).

Of course, this doesn't mean that you'll never have problem employees who don't respond to this positive approach. As we discussed in Chapter 3, those people who can't or won't perform must be dealt with. However, for the most part, the Pygmalion Effect works. If you're interested in learning more about this phenomenon, just type "Pygmalion Effect" into your search engine and you will discover a wealth of information.

The effects of having confidence in yourself and others are possibly more far reaching than just the Pygmalion Effect, but at least this gives some sound basis for the premise that you should approach your job with confidence— confidence in yourself, the success of your endeavors, and the people you lead and work with.

We've established that confidence breeds confidence and promotes success. Unfortunately the opposite is also true if you don't feel confident. Perhaps sharp criticism from your boss has stripped you of your confidence. Or maybe you aren't ready. You haven't prepared your speaking notes. Or you haven't completed that last design calculation. Or you haven't thought through all the possible commercial risks. Or you haven't run that last economics case. The list goes on.

In circumstances like those, you have to dig down and find a way to bolster your confidence when you need it. I recently watched an international track and field competition on television. A high jumper stood, focused on the bar as he prepared to start his run. He suddenly held his arms out to his sides and began clapping his hands over his head in the same cadence he would use to begin his approach to the bar. The home crowd picked up the cadence and clapped as he started his run. He cleared the bar and the crowd exploded in a cheer. This high jumper had found a way to lift his confidence.

That particular approach won't work for you in the office, but it illustrates the technique. You have to find some personal way to psych yourself up when you need it. I'm convinced that energy follows thought. If you focus on the task at hand and visualize the desired successful outcome, it's more likely to occur.

YOU HAVE A SOUND BASIS FOR YOUR CONFIDENCE

With your education and sound interpersonal skills as a foundation, along with understanding the principles and practicing the strategies we've covered in this book, you have an excellent basis for confidence in yourself and your abilities. If you continue to explore, understand, and apply this practical knowledge, until it is an innate part of who you are, you'll have confidence as your traveling companion for the rest of your career.

THE FOUNDATION: EDUCATION AND INTERPERSONAL SKILLS

Education provides the basic tools of the engineering trade. Supposedly, the better the school, the better the education, but in these times with so many excellent universities to attend, most of you will be equipped with sufficient engineering skills. And—independent of which school you attended—you will have gained an essential frame of reference for most aspects of engineering, technology, and research. You'll have had enough small projects under your belt to understand how to approach them and how to cooperate with others as a team. If you've worked in the business world during college, you'll have a frame of reference for that too. Another by-product of education is the set of analytical skills you've developed when confronted by homework assignments, tests, labs, and everyday academic and social situations.

Meanwhile, depending on how you've approached your studies, your basic work ethic will have started to form and mature. If you've taken part in extracurricular activities or student governmental positions, you may have learned organizational and interpersonal skills—perhaps the hard way. You'll at least have some sense of how well you communicate and get along with other people.

With your education and all these experiences at your command, you can confidently enter the workplace. There you can use the set of skills, strategies, and options for further development that we've discussed earlier in this book to help you succeed.

Let's review everything and pull it together into one big picture.

To Get Started in the Workplace

In Chapter 1, I repeated these three rules that were passed on to me early in my career and stood the test of time thereafter:

- Find out who your boss is (or bosses are).
- Find out what he or she wants (or they want).
- Do it!

In the process of explaining "what your boss wants" in Chapter 1, I introduced the concept of Management by Objectives, without naming it. Often called MBO, this management approach is used successfully in North America and in many (but not all) societies throughout the world. Peter F. Drucker, the man who first coined the phrase "Management by Objectives," gives an excellent discussion of MBO in Chapter 34 of his book, *Management: Tasks, Responsibilities and Practices* (Drucker, 1993). I thought it noteworthy that he titled the chapter "Management by Objectives and Self-Control," which emphasizes that

- objectives should be aligned across the organization, and
- managers and employees alike *want to contribute and will take the actions necessary to achieve the objective task without being directed from above.*

Fortunately, when it comes to *doing it*, there are some basic duties that most project engineers have in common.

The Project Engineer's Basic Duties

Project engineers' primary role is to exercise total responsibility for everything that has to do with their areas. They know to ask questions until they understand what their areas are and what the boundaries are. Their fundamental duties are the following:

Duties of a Project Engineer

- Plan and control the basic work.
- Lead safety.
- Identify, assess, and mitigate risks.
- Achieve quality standards.
- Control schedule within the plan.
- Control costs within the budget.
- Control interfaces.
- Manage changes.
- Solve problems and commercial issues.
- Lead the effort.

Those cover the full scope of project engineering but a specific job may not have all of those responsibilities. For example, systems engineers may not have the cost and schedule responsibilities. They are often responsible for coordinating the design, development, and manufacture of their system, while a project management group looks after the project control function for the whole project. You can use this list of duties when asking your questions about the scope of your responsibilities, and you can make your own list tailored to your specific job—deleting the irrelevant ones and possibly adding some.

Project engineers must balance the priorities of safety, quality, cost, and schedule. Safety can never be compromised for those other goals. Quality must be adequate for the purpose. Cost and schedule have to be balanced to achieve the project objectives. You can't work exclusively on saving money and miss a scheduled milestone that results in your organization incurring a large penalty payment.

Management of the interfaces with the rest of the project and the outside is crucial. Here is a basic process that is useful in resolving the interface issues that are bound to occur:

Useful Interface Process

- Define the interface and develop a plan for drawings and documents that need to be exchanged and agreed on.
- Identify issues (mismatches, clashes, lack of definition, schedule problems).
- Agree, *in writing,* to
  - the tasks needed to resolve the issues,
  - responsibilities for the various tasks, and
  - a date for completing each task or providing information to the other party.

Solving technical, project execution, and commercial problems is a part of the daily routine for project engineers. Problems come from design changes, lack of communication, and a host of other root causes: a careful reading of Chapter 5 will illuminate how thorny the problems can become and how to resourcefully handle them.

You can focus the problem-solving effort with a three-step, seven-substep approach:

Problem Solving

---

1. Don't avoid the problem.
2. Focus on the problem, and if necessary get experts involved. These experts should be
   - people who have essential information or expertise to contribute, or
   - people who will need to be committed to the solution when it's time to implement it.

3. Resolve the problem and document the solution. This usually takes
    the form of one or more of the following:
    ○ A formal change (within the project's management of change process)
    ○ A change to design drawings and documents
    ○ A change to one or more contracts
    ○ A change to the purchase orders involved
    ○ Possibly some other kind of documentation, such as an interface
       agreement, which will eventually work its way into the project
       drawings and documents

Finally, as you fulfill your commitment of being totally responsible for everything that has to do with your area, mentally prepare yourself to lead.

HOW TO MANAGE TASKS AND PEOPLE

A crucial part of a project engineer's job is managing tasks and people. This is what differentiates a project engineer from other more technically oriented engineers and gives him or her the opportunity to exert greater influence over the work. As pointed out in our "Crash Course in Management" (Chapter 3), the starting point is to think and act like a manager—like your role models who are already managers.

The functions of management from Chapter 3 outline the scope of managerial responsibility. You'll note that we have picked up an additional function on the people side—cross-cultural communication skills—from Chapter 8:

Functions of Management

Task-oriented side	People-oriented side
• Plan	• Motivate
• Organize	• Build a team
• Delegate	• Develop people
• Control	• Use cross-cultural
• Integrate	communication skills
• Measure	
• Improve	

Not all of these functions will apply in every situation, but they will serve as a checklist to assure you that you have covered all the bases from a managerial point of view. When you approach a new task, think about plan, organize, delegate, control, integrate, measure, improve. Think through each of those functions and decide how you will handle them.

Input from your team will improve the plan and build your team members' commitment to it. Remember, after your early years you'll seldom be flying solo. Instead you'll be leading a team.

On projects, the planning is critical (remember "Plan the work and work the plan"?). The project engineer's planning process, covers most of the planning steps needed for both small and large jobs:

Project Engineer's Planning Process

---

1. Set objectives—make sure you have defined specifically what is to be done and when it is needed.
2. Establish that the work is feasible and adequately defined.
3. Check the input data and requirements.
4. Break the work down into activities and assign responsibilities to team members. Bring in experts as needed.
5. Assign a duration to each activity and establish the dependency between activities (for example, which activities must be finished before others are started).
6. Estimate the work-hours and cost to complete each activity.
7. Finally, develop a baseline cost and schedule estimate that is aligned with the project's cost and schedule estimate.

Organizing is integrated into the above planning process and involves getting the right people on your team and assigning responsibilities to each person. The plan will be your basis for controlling, measuring, and improving the work.

When you get past the planning and into the implementation, recall that prioritization and shameless delegation of the tasks are essential if you want to keep your head above water. You will spend most of your time directing or controlling the work of your team and integrating that work with the activities on the rest of the project. You'll also solve problems, keep up with the schedule and budget, and monitor all the other project engineers' duties.

While you are addressing the task side, keep in mind the people side: motivation, team building, personnel development, and cross-cultural communication when it's relevant. Effectively managing your team will require your best interpersonal skills. We won't repeat Chapters 3 and 6 here, but will simply state that influencing people is at the heart of successfully managing your team. As I wrote in Chapter 3 at the beginning of the "people side" discussion:

The most effective managers are able to balance accomplishing their tasks with taking care of their people. They keep people informed. They get people engaged in the work and onto the playing field. They create an environment that allows people to participate in decision making and be empowered to take action. They lead them through difficult times with confidence and take pressure off during stressful situations.

When things get chaotic and stressful, remember Henry Mintzberg's take on management in "The Managers Job: Folklore and Fact" (Mintzberg, 1990). It may help you lift your spirits up a notch and manage more effectively.

Whenever you need to do a quick self-assessment of how you're managing the job, ask yourself the questions at the end of Chapter 3. I recommend that you ask those questions early and often, at least until management is second nature for you.

How Projects Work

If you are assigned to a large project, make sure that early on you review "How Projects Work" in Chapter 4. It's a concise summary of the major phases, activities, and considerations for planning and implementing large projects. The overall project framework is repeated here for reference in Figure 10.1.

Prior to approval, the project is planned, evaluated, and defined. Front-end engineering design (FEED) develops the engineering basis for the contracts. The client prepares bid documents, issues them, and drafts the contracts. Contractors line up their partners and suppliers, prepare their bid proposals, and try to mitigate their risks. Project Execution Plans (PEPs) and cost estimates are prepared by all the major parties involved.

Once the project is approved by the client, it moves into the implementation phase. Contracts are awarded, and the successful contractors are brought on board to carry out engineering, procurement, manufacturing, construction, commissioning, and sometimes operations. The management of those major

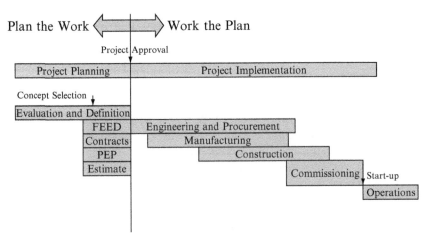

Figure 10.1   Overall project framework.

phases, the transition of deliverables between them, and the interfaces between the main parties require skillful project management on everyone's part—clients, contractors, and manufacturers.

The concepts in Chapter 4 provide valuable background. It's a frame of reference for how the project around you functions and how your area fits in. And it provides signposts to warn of what's ahead so you can plan and react. If you're a design project engineer, you'll know to interact closely with procurement and prepare to deliver the drawings, documents, and data registers to construction on time. If you're a construction project engineer, you will be aware that deliveries of design documentation and equipment are often late, and you will be prepared to start pressing for them at the first signs of delay. You will also look ahead by reviewing the commissioning plans and coordinating the precommissioning of the systems within your area to meet those plans.

JOB EXPERIENCE FROM THE CASE STUDY

The case study in Chapter 5 is intended to be food for thought. Store these situations in your experience bank and draw them out when needed. Pondering the insights from the case study will give you some familiarity with the real situations you will encounter, which leads to self-assurance.

Keep Edgar's Engineering Team Process handy when you are starting to form any team, large or small:

Engineering Team Process

- Communicate:
  - Early
  - Often
  - Informally
  - Candidly
- Keep our credibility high.
- Use benchmark data to plan, schedule, and budget.
- Be involved—hands-on project engineering and technical leadership.
- Don't say, "No." Say, "Let me work on a way to help."
- Develop effective relationships (project, client, subsuppliers, and management).
- Expand your role.

KNOWING WHAT IT TAKES TO GET AHEAD

For young engineers, Chapter 6 is a road map for how to succeed in their careers. It gives options and strategies for building competence and the other skills it takes. However, you are the one who must decide where you want to go and what route to take. While on the journey, you will select which turns you will make and how much time to spend in each place.

We went beyond the three elements of Business World 101—competence, office politics, and social skills—and explored the facets of personal effectiveness, business judgment, and a strategy for competing with your peers. Just knowing about these concepts gives you an advantage over your contemporaries and is a source of self-assurance.

Personal Effectiveness

Personal effectiveness is a series of nontechnical skills that relate to *how* young project engineers approach their jobs. These will amplify performance and promote success.

Getting Organized

The starting point is maintaining a few simple processes to keep your work organized: a task list, a calendar, handling mail only once, optimizing the use of email, keeping a journal of meeting results, and using a call-up file to jog your memory. In this electronic age, new technologies are constantly emerging to facilitate getting organized.

Initiative

Top performers have an attitude called initiative that sets them apart:

- They handle their job without needing much guidance.
- They anticipate what they and their bosses need.
- They expand their roles.

Communication

More than any other factor, communication skills promote effectiveness. Everything from important conversations, to writing, to presentations, to simply listening will reveal the engineer's competence to the rest of the organization. The ability to sell ideas and stay focused on business results is perhaps the key factor in being effective. Fortunately, these skills can be developed through training.

Effective Meetings

One of the best ways to achieve results—your results—is to conduct effective meetings. Every meeting should have the following:

- Purpose
- Agenda
- Wrap-up session that clearly sets out the meeting's conclusions

If presenting a proposal to management, it should have a recommendation that can be acted upon. The meeting's chair should facilitate the discussion to create a supportive environment, if possible.

Finishing the Job

Engineers—particularly those who have strong technical skills—sometimes have difficulty completing their assignments. If you find this to be a problem, seek the help of your mentor or a trusted friend in the organization.

Teamwork and Leadership

Sound interpersonal interactions form the basis for teamwork and leadership. Being an effective team member is an important first step. When you move into a leadership position, you'll have to balance

- getting the job done well,
- taking care of the people, and
- managing the situation.

Power and Influence

As a project engineer, you should recognize that you have three types of power available to leverage your abilities: expert power, informational power, and personality power. Use them wisely. Your goal is to influence others over the long term, and these forms of power are a few of the tools at your disposal.

Essential Training

Personal effectiveness can be learned! One of the best things you can do to enhance your performance and career is to take some essential training early to improve your time management, presentation, writing, teamwork, interpersonal, and supervisory skills.

Business Judgment

You possibly hadn't thought of business judgment as a key part of competence, but I can assure you it is. Business judgment starts with learning about the organization with the purpose of understanding the structure of the company, its business objectives, its main players, how the management thinks, and how the company works. Strive to understand your own organization and then branch out to other units. You can objectively analyze the situations you are involved in by asking the following questions:

- What worked?
- What didn't work?
- Why?

Over time, your business judgment will mature, and you will intuit what should be done and how to sell your ideas.

Performance Evaluations and the Competition

Keeping a healthy attitude toward competition in the workplace is your best approach. You'll need the help of peers and friends to get your job done, so you don't want to be regarded as a backstabber. By concentrating on the things that matter—results, competence, effectiveness, building business judgment, making friends among your peers, cultivating management patrons, and getting along with others—you'll compete in an ethical, rewarding, and successful way!

As you progress from day to day and decade to decade, make sure you don't forget Business 101. These are things you learn early but always need to know: competence, office politics, and social skills.

The trickiest, and the most likely to trip you up years into a successful career, is the term in the middle: office politics.

Office Politics

As with competing, it's wise to concentrate on the positive side of office politics. Building a network of allies among your organization's management, peers, and subordinates is one of the best defenses against political maneuvering by others. First, the network will give forewarning of problems. Second, management patrons are particularly valuable since they will be involved in private conversations about these matters and can influence decisions in your favor. Also getting the right kind of exposure in the organization will create a positive image of you among the management that will help you weather political storms you didn't even know were brewing. And, if you do end up in an office that's steeped in political gaming, you can fall back on Patricia Young's (2003) survival plan in Chapter 6.

Never let yourself be ensnared by dirty politics.

The Business 101 skill set isn't generally taught in most engineering universities, but they will measurably improve your performance and help you chart a sound course for your career. In addition, they'll build your self-reliance.

Those nontechnical skills that can get you ahead in positive times will save your bacon in negative situations: when unethical conduct is either a temptation to you personally, or something those around you are doing. Chapter 7 dealt with the unacceptable conduct that can get you fired.

An Understanding of Acceptable Business Conduct

Ethical business conduct is good business. It avoids incidents and scandals that can sink a career or even a company. Also, as many chief executive officers claim, ethical conduct gives a competitive advantage in the market place over the long haul. Just understanding the importance of business ethics and complying with your company's code of business conduct will give you a sense of strength and self-confidence.

When you find yourself in a situation that requires ethical judgments and you have to draw the line, you will follow the high road and take the following honest, transparent approach:

An Approach to Resolving Business Conduct Situations

- Know the relevant laws, regulations, and your organization's business ethics policies.
- Evaluate the situation.
- Discuss it with others and seek advice.
- Try to isolate what's illegal or unethical and avoid it.
- Treat the situation with openness and honesty.
- Then decide.

This approach will keep you on your toes at all times. You may need it most, and find it hardest to apply, at the point that everything else becomes harder too: when you are assigned to a foreign country. But using the guidance in Chapter 8, you can turn your international experience into a plus—a memorable part of your life and an ingredient in your professional confidence.

How to Approach Working Internationally

You can draw self-assurance from the knowledge that you are aware of cultural differences and their impact on international business. You're aware of culture shock and how to deal with it in your reactions when you first arrive at a foreign assignment. And once you get your feet on the ground in the new country, you have an international tool kit:

Tools to Deal with Cross-Cultural Situations

- Awareness of and sensitivity to intercultural issues
- A commitment to understand what the other person means
- A set of cross-cultural communication skills
- A basic approach that will help you deal with cultural differences:
  - Setting the right climate and gaining rapport with your counterparts
  - Establishing a framework to collaborate in
  - Understanding how to create a dialog, resolve differences, and reach agreement on the issues of the day

---

Whether you are in your home country and culture or abroad, one of the smartest things you can do is get good advice. And, of course, listen to it!

SOUND ADVICE FROM PROJECT PROFESSIONALS

There is no one way to be a project engineer—jobs are different, situations are different, and people are different. A person needs a set of skills and good judgment to succeed. Chapter 9 offers you the collective judgment of experienced project managers and professionals, which you can use to develop your own. They are responses to the question, "What advice would you give a new project engineer?" Draw on those that seem to fit, especially those from the young project engineers.

As you go through professional life, you will find that most good people actively enjoy giving advice to younger or less experienced colleagues. The advice in Chapter 9 represents a fraction of what's out there in the engineering world for you. You won't regret getting and internalizing advice from the people ahead of you.

This, by the way, is what will guide you when going above and beyond the limits of the ground I could cover and the situations I could predict in this book. Your future is more open ended than you yourself can predict. If you have the interest to ask for advice, the humility to listen to it, and the well-grounded confidence to act on it, then you can face the future with a great deal of well-founded optimism and zest.

THE OPPORTUNITIES ARE BOUNDLESS

Does our global society need large projects to build and renew our infrastructure? Do we need projects to discover, develop, and produce more oil and gas, or invent and create alternative energy systems that will serve us for decades and centuries to come? Will there be projects to improve and create information technology systems or related equipment, and cope with the increased technical complexity? Will we need to develop and manufacture the systems to keep our society safe? Will companies manufacture the transportation of the future? I'm sure we will. I'm also certain there will be project engineers that integrate, coordinate, and lead the efforts of others within their areas of responsibilities and beyond.

Approach your job with confidence and commitment and then things will start to fall in place. Take care of the job and the people. Conduct your business with high standards of ethical conduct. Be sensitive to the ideas and concerns of people from other cultures. Learn from those around you and those who have preceded you.

You have the tools in your toolbox to launch your career and to succeed. *"Begin it!"*

REFERENCES

Accel-Team.com, "The Self-fulfilling Prophecy or Pygmalion Effect," http://www.accel-team.com/pygmalion/prophecy_01 and 05 (2005).

Drucker, P. F., *Management: Tasks, Responsibilities and Practices* (Harper Business, New York, 1993), pp. 430–432.

Mintzberg, H., "The Manager's Job: Folklore and Fact," *Harvard Business Review on Leadership* (The Harvard Business School Press, Boston, 1990), pp. 4, 12–21, 29–31.

Shaw, B., *Pygmalion* (Brentano, New York, 1916); Bartleby.com, www.bartleby.com/138/ (1999).

Young, P., "When Good Isn't Good Enough," *The Globe and Mail* (Calgary, Canada, September 17, 2003), pp. C1, 6.

Glossary

AFC approved for construction

AFD approved for design

Antitrust laws laws enacted to curb noncompetitive behavior by private companies

Area designated part of a facility, portion of manufactured equipment, system, group of systems, purchase order, work package, or set of related tasks for which a project engineer is responsible; has specified boundaries and interfaces with other parties

Area engineering phase of project, detailed engineering when areas are engineered in sufficient detail for bidding and later for construction

As-built documentation final drawings, documents, and data that reflect the finally constructed project facility (or sometimes manufactured item)

Back-to-back contract subcontract that transfers all the terms, conditions, responsibilities, and risks of the main contractor to the subcontractor

Benchmarking comparison of a cost, a schedule, or other project estimate with actual, similar data from previous projects

Bid-check estimate estimate for a specific contract or purchase order that is prepared prior to receipt of the bids to judge the reasonableness of the quoted prices

Bidding process events from a decision to contract for work through award of the contract; includes preparation of the invitation to bid, selecting the bid list, sending out invitations to bidders, bidders' preparation and submittal of proposals, preparation of contract documents, evaluation of the bidders' proposals, negotiations with the successful bidder, and signing of the contract; same process applies to purchase orders; also known as *tendering process*

Bid package documentation sent out with the invitation to bid

Buyer commercial person who prepares, negotiates, and administers a purchase order for a client or contractor; works closely with a package engineer

Change order variation to a contract or purchase order that adjusts the scope of work, contract price, or other aspects of the contract or purchase order

Clash interference of two design elements, e.g., a major pipe interferes with a main structural member and must be rerouted

Client owner or operator of the project or venture that is overall responsible for specifying, managing, and procuring goods and services procured under a contract or purchase order

Commissioning also called *systems completion;* phased completion, testing, and functioning of systems or subsystems prior to start-up and operations

Confidence   one of a project engineer's most valuable assets

Conflict of interest   exists when an individual has or appears to have a personal interest that is contrary to his or her organization's interests

Construction   planning, fabrication engineering, procuring, prefabricating, fabricating, erecting, assembling, transporting, and installing a project facility

Contingency   additional cost added to an estimate to cover unknown risks

Contract   written agreement to provide goods and services

Contract—lump sum   contract in which the contractor bids a total fixed price for completing the work; generally paid at payment milestones spread throughout the duration of the contract

Contract—reimbursable   contract in which the contractor is reimbursed for all actual costs and is paid a fee for the performance of the work

Contract—schedule of rates   contract in which the contractor bids a list of rates for performing elements of the job (e.g., cost for installing and testing piping, cost per fabricated ton of steel, cost for the length of a given type of welding); number of rates can vary widely depending on the contract; also called a *unit rate contract*

Contracting   agreeing on the provision of goods and services through a contractual arrangement

Contracting strategy   plan for dividing the project work into contracts; the client's contracting strategy drives the contractors' contracting strategies; see Figures 4.2, 4.3, 4.4

Contractor   firm, joint venture, or consortium that is contractually responsible to provide goods and services

Control cost estimate   project cost estimate that forms the baseline for controlling project costs; developed early in detailed engineering or prior to project approval; may be the basis for funding the project

Control schedule   project schedule that forms the baseline from which progress is measured; generally developed prior to project approval and contract awards or in the early stages of detailed engineering

Corrective action   deficiency in a project management system or procedure, usually found by a quality audit, that must be corrected by the project management

Cost allowance   cost added to an activity or overall estimate to cover known risks; see also *risk-weighted cost allowance*

Data registers   on projects: large electronic databases that define equipment to be bought, materials to be ordered, mechanical completion status, commissioning status, and more

Design basis   collection of information required to design a facility or other design object; can consist of environmental data, design criteria, functional requirements, standards (design, safety, quality, regulatory), a conceptual design, and other types of information and data

Design change   significant change to a concept (during project planning) or a change to the design (during project implementation); the project must manage change

E&P contractor   firm, joint venture, or consortium that is contractually responsible to provide engineering and procurement services; also a contracting strategy in which the client decides to use an E&P contractor for closer control over engineering and procurement than an EPC strategy would give

El dinero   Spanish for "the money"

Engineering and procurement   integrated project phase to design, specify, purchase, and contract for everything necessary to execute and initially operate a project

EPC   engineering, procurement, and construction; generally refers to a contract or contracting strategy

EPCM   engineering, procurement, and construction management; generally refers to a contracting strategy that gives the client more control than an EPC strategy over construction

FEED   front-end engineering design; part of the definition phase of a project

Function testing   testing of a piece of equipment by running it

Functions of management   task side: plan, organize, delegate, control, integrate, measure, improve; people side: motivate, build a team, develop people, use cross-cultural communication skills

Getting forward   slang for ensuring that a deliverable is accomplished on time, e.g., "responsible for getting forward the process flow diagrams"

Hazard register or risk register   list of project risks and hazards that results from a HAZID, HAZOP, or other risk assessments; risks are generally arranged from most severe at the top; sometimes called a *risk register*

HAZID   organized group session to identify hazards and risks to a project

HAZOP   hazard and operability study to improve the safety aspects of a design; consists of a review, usually during detailed engineering, that involves

personnel with engineering, safety, and operational knowledge and experience

Hold points   places in the manufacturing process where the customer requires the work to stop while some specific action (like an inspection or test) is taken

Holds   areas on a drawing (usually surrounded by a cloud) that are lacking necessary information to carry out the work

IDC   interdisciplinary checks; design quality measure to check design outputs; assures that discipline engineers have an opportunity to check the impact of other engineers' design work on their own

IFC   issued for construction

IFD   issued for design

Inspection and test plans   measurements, tests, inspections, and verifications planned to manage the quality of constructed and manufactured products

Interfaces   internal or external boundaries with other parties at which information, designs, activities, and deliverables must be managed

ISO   International Organization for Standardization

JSA   job safety analysis; formal or informal assessment of risks and mitigation measures done by a work crew prior to starting a task

Leadership   activity of influencing people to strive willingly for group objectives (Hersey and Blanchard, 1982)

Local content   requirements by host countries on international projects that certain amounts of the work be done in-country

Long-lead equipment   equipment (or materials) with a long delivery time; the project must order this equipment before project approval in order to receive it on time; generally requires a special funding request

Loop testing   continuity testing of instrument or electrical circuits to verify mechanical completion

Management   working with and through individuals and groups to accomplish organizational goals (Hersey and Blanchard, 1982)

Management by Objectives (MBO)   system of management based on establishment of a cascading set of increasingly detailed objectives, which are mutually agreed on between bosses and subordinates; focuses the work on overall organizational goals; under the right management style, all levels work in a self-controlling manner toward these aligned goals and objectives, which should be achievable and measurable, i.e., specific results delivered at an agreed time

Management of change (MOC), or change management   project control process to ensure the orderly coordination, control, approval, and implementation of project changes (including changes to the design)

Manufacturing   production of equipment and materials; consists of planning, engineering, procuring, developing, producing, testing, and delivery to the customer

Materials   purchased items such as wire, pipe, steel, and more

Mechanical completion   documented construction status for an area or system that indicates that it is completely built, hooked up, and tested so that it is safe to be commissioned or operated on a trial basis.

Mega-project   project that is extremely large with a total installed cost of approximately $1 billion U.S. or more

Milestone   project date specified in a contract at which certain deliverables are due or certain events shall happen

Model contract or PO   template used by contract (or PO) formulation teams to produce a specific contract (or PO) document; templates help ensure uniformity and consistency while allowing for necessary variations

Nonconformance   deficiency or deviation from requirements (found by any means) in a product that must be corrected by the project management

Organization chart   see Figure 3.2

P&ID   piping and instrumentation diagram indicating the essential details of a process or subprocess

Package engineer   project engineer responsible for an engineering work package, a purchase order to be delivered, or a portion of a manufacturer's scope of supply

Patrons   management sponsors that can help accelerate an individual's career

PEP (Project Execution Plan)   comprehensive implementation plan that covers all aspects of the client's, contractor's, vendor's, or other responsible party's work on a given project

Personal effectiveness   how well an individual performs his or her job; see Chapter 6

Power   potential for influence (Hersey and Blanchard, 1982)

Present value profit   value of a future net cash flow stream discounted back to its present total value; the annual rate used to discount each year's cash flow back to the present is usually a fixed percentage such as 10% or 15%; PV10 would be the net profit discounted back at 10%

Procurement   purchasing and contracting

Project—new project engineer's view torrent of activity, rushing along, sweeping away everything in its path

Project—project manager's view structured set of related plans and activities, executed in a particular order, to accomplish specified objectives

Project approval decision to implement a project and fund it

Project engineer engineer who is responsible for a system or area of a project and coordinates the work of others to get the work completed

Project implementation also called *project execution*; project phases after approval: engineering and procurement, manufacturing, construction, and commissioning and start-up

Project planning evaluation of a project's viability measured against specified criteria and weighed with business judgment, selection of a project design concept, and definition of that concept

Proprietary information information that a company considers to give it a competitive advantage

Prototyping manufacturing one of a series of products ahead of full production

PSB planning, scheduling, and budgeting

Punch list list of deficiencies that must be corrected; results from an inspection near the end of construction, manufacturing, or other such activities; deficiencies must be completed before the product is accepted

Purchase order (PO) written agreement to provide equipment, materials, and sometimes services; generally less comprehensive than a contract

Purchasing buying of goods (and sometimes services) through a purchase order

Qualitative risk assessment risk assessment based on the experience and judgment of project personnel to determine the probability and severity of the consequences of risk events

Quality management process-based approach that converts client requirements and inputs into an acceptable product through a cycle of management responsibilities, allocation of resources, product realization activities and controls, measurements, feedback, and process improvement

Quality surveillance and expediting oversight of purchase orders for equipment and materials by procurement and engineering personnel; generally involves assessment of the vendor's quality systems, conformance with technical requirements, and ensuring on time delivery

Quantitative risk analysis (or assessment), QRA structured approach of assessing project risks that calculates risk probabilities and levels, then measures them against predetermined criteria; generally requires experts with access to historical risk data to perform this type of risk assessment

Responsibility matrix see Table 3.1

RFP request for proposal; invitation to bid on a contract

RFQ request for quotation; invitation to submit a price for a purchased item

Risk matrix see Figure 2.1

Risk money cost allowance added to a proposal by the contractor to cover known risks

Risk-weighted cost allowance amount added to the control cost estimate to allow for known risks; the cost associated with each known risk is multiplied by its probability of occurrence and summed to calculate the risk-weighted allowance

Root cause basic or systemic cause of an incident or accident (e.g., faulty procedure, not following procedure)

Schedule—critical path activities from beginning to end on a schedule that determine its total duration; a delay in critical path activities will delay the completion date of the project, unless a recovery plan can be developed

Schedule—front line line drawn vertically on a bar chart schedule to indicate the degree of completion of each activity (bar); if the front line for a particular activity is to the right of the present date, that activity is ahead of schedule and vice versa

Schedule—level 4 generally a schedule of detailed activities; definition varies from organization to organization

Schedule—level 5 generally a schedule that includes every document, drawing, data sheet, report, specification, work package, and other deliverables to be completed by the project

Schedule reserve allowance (usually several months) added by project management or higher authority near the end of a schedule to account for known or unknown risks

Scope of supply listing of what will be delivered under a contract or purchase order (list of deliverables to be furnished to the client)

Scope of work part of a contract that describes the work to be undertaken by the contractor

Shop drawings engineering drawings that give the manufacturer's shop personnel the details to fabricate equipment and other items

Short-listing reduction in the number of options to two or three; can refer to the selection of a few bidders that will be evaluated in more detail before a final selection is made; also refers to a reduction in the number of concepts prior to final concept selection

Site queries questions from the construction sites back to engineering to clarify design matters; there is a system for this on larger projects

Six sigma quality management process-based, quality management approach that seeks to reduce unacceptable variations (defects) in the process to below 3.4 defects per million by process improvement, process management, and process design or redesign

Specification project document that prescribes (usually in detail) how something is to be engineered, manufactured, constructed, or used; e.g., design specification, compressor specification, paint specification, steel specification, piping specification, flange torquing specification, heavy lifting specification

Specs slang for specifications

Start-up phased process of bringing a process or plant into full or partial operation

Subcontract contractual agreement between a contractor and another contractor to provide goods and services for a portion of the originally contracted work

Subproject part of a large project that operates as a project within a project because of its size or importance

Supplier vendor or contractor

Systems completion see *commissioning*; sometimes includes construction work that could not be or was not completed in earlier phases

Systems engineer project engineer responsible for a system or group of systems

Systems engineering initial phase of project detailed engineering or manufacturing when systems are defined, engineered, and specified

Technical requisition technical requirements for a purchase order prepared by the engineering team

Total area responsibility project engineer's mind-set that he or she is completely responsible for everything that has to do with his or her area

Validation overall demonstration that a product functions as designed

Variation order same as *change order*; some projects use variation order for contract changes and change order for purchase order changes

Vendor party to a purchase order who agrees to provide the goods and services

Vendor data engineering data required from manufacturers; it is necessary to engineer interfaces and complete the details or project systems and area engineering

Verification confirmation that a work product meets its specified requirements

WBS work breakdown structure; detailed listing of every significant part of the contractual work; includes components, subcomponents, work packages, and other project deliverables

Widget humorous name for a manufactured product

Witness points activities in the manufacturing process that the customer wants to observe

Work-arounds approach to get around a problem or mitigate risks; synonymous with contingency plans

Workface slang for the level in an organization where work actually is done; analogous to the "coal-face" in a coal mine

Work-hour hour worked by a man or woman

Index

Printed and bound by CPI Group (UK) Ltd, Croydon, CR0 4YY

03/10/2024

01040434-0003